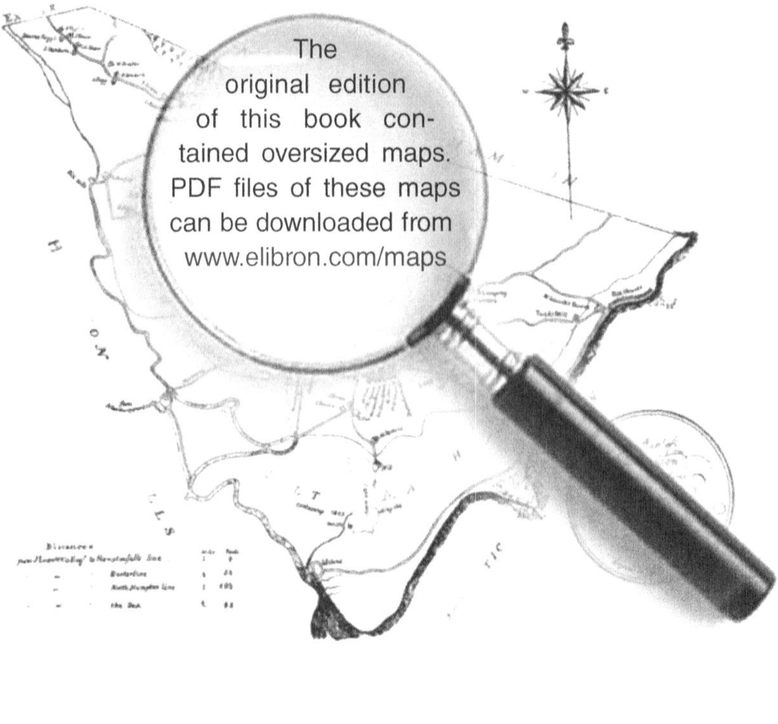
The original edition of this book contained oversized maps. PDF files of these maps can be downloaded from www.elibron.com/maps

Francis Edward Clark

THE GREAT SIBERIAN RAILWAY

Elibron Classics
www.elibron.com

Elibron Classics series.

ISBN 1-4021-8513-8 (paperback)
ISBN 1-4021-0859-1 (hardcover)

This Elibron Classics Replica Edition is an unabridged facsimile of the edition published in 1904 by S. W. Partridge & Co., London.

FRANCIS E. CLARK

THE

GREAT SIBERIAN RAILWAY

WHAT I SAW ON MY JOURNEY

BY

DR. F. E. CLARK

WITH SIXTY-FIVE ILLUSTRATIONS
FROM PHOTOGRAPHS AND A MAP

London
S. W. PARTRIDGE AND CO.
8 & 9, Paternoster Row
1904

CONTENTS

ILLUSTRATIONS

Illustrations

Illustrations

Illustrations

BY WAY OF INTRODUCTION

THIS volume describes the last six weeks of a long journey undertaken in the interests of the Young People's Society of Christian Endeavor.

Mrs. Clark and myself having been invited to attend various conventions of the society in Japan and China, we started for the far East, accompanied by our little son Harold, a boy of twelve years, in the early days of January, 1900.

The conventions were, beyond our expectations, large in numbers, enthusiastic in interest, and important in their results; but they cannot be described in this volume.

Knowing that a new all-steam route around the world was to be opened by the completion of the Trans-Siberian Railway to the head of navigation on the Shilka River, and having important engagements at the World's Christian Endeavor Convention, which was to meet in London in July, we determined not to follow the beaten tracks around the world which we had before taken, but to attempt this new route, which binds Asia and Europe, the Pacific with the Atlantic.

The last meetings of the series we had gone to the far East to attend were held in north China, in Tientsin, Peking, Tungchow, and Poatingfu, only a few days before the terrible uprising which bids fair to make such momentous changes in the map of the world.

The meetings being over and our object in visiting China being accomplished, we left Taku, as it proved, by the last steamer that sailed for Corea before the bombardment of the forts of Taku and the siege of Tientsin.

For nearly two weeks we coasted around the peninsula of Corea, stopping at Chemulpo, Mokpo, Fusan, and Wonsan, or Gensan, as the last-named port is indifferently called.

At Fusan, at the southern extremity of Corea, we spent four days waiting for a steamer to take us up the coast, and here, in soil already prepared, the first Christian Endeavor Society of Corea was planted in a church of the Australian Presbyterian Mission.

At length we reached Vladivostock, and our long journey across the Siberian plains by rail and river began.

At Taku our party had been pleasantly enlarged by the addition of Miss Anna Northend Benjamin, of New York, a well-known lecturer and newspaper correspondent, on her way home from the Philippines.

At Wonsan, in Corea, Mr. J. M. Jordan, British Consul-General to Corea, and Mr. Burn Murdock, a mining expert, became our fellow-passengers.

At Khabaroffsk Mr. J. C. Smith, a leading merchant of Nagasaki, who was accompanied by Mrs. Smith, took the same steamer; also Baron Hartogensis, of Germany, and Herr Ritter, Swiss Consul-General to Japan, all of whom, with one or two other foreigners (non-Russians), afterwards formed themselves into the "Never Again Society," to whom this book is dedicated.

Interest is perhaps added to this journey from the fact that we were the first Americans, and in all probability the first foreigners, to go around the world by the new route. The trans-Siberian all-steam route had been opened but a few days when we took passage at Vladivostock, as I have explained in an early chapter, and none of the few who had preceded us across Siberia by this route had completed the circuit of the globe.

If this fact is borne in mind, it will explain some of the difficulties inevitably incidental to a new and untried route.

I wish gratefully to acknowledge the courtesy of their Excellencies the Russian Ministers Plenipotentiary to Japan and China, and to General Tchitchagoff, the

Governor of Vladivostock and the Primorskaia, for their kind assistance in a journey which, were it not for their aid, would have been far more difficult and tedious.

It is also fair to add that already, in the minds of some at least of the "Never Again Society," the discomforts and delays of the journey have faded into the unremembered Past, which so kindly and so quickly touches our little woes with the finger of oblivion, while its novelties and pleasures stand out in ever-bolder and happier relief. If the above-mentioned society were canvassed to-day, it would not be at all surprising if its members should unanimously resolve themselves into a

"Some Time Again Society."

BOSTON, *February* 2, 1901.

A NEW WAY AROUND AN OLD WORLD

I

FROM NEW YORK TO VLADIVOSTOCK

THE completion of a new way around this old world is a matter of no little interest and moment. Such a new way has just been opened, and it was my fortune to be one of the first five passengers who took this journey.

On the 28th of December, 1899, the last rail was laid on the trans-Baikal section of the Siberian Railway, which for the present will stop at Stretinsk, and there connect with the steamers on the Shilka, a branch of the Amour River, for Khabaroffsk, where again railway communication is resumed for a further journey of five hundred miles to Vladivostock, on the Sea of Japan. Thus across all of Russia's vast domain in Europe and Asia it is possible to go by steam, a distance of considerably more than six thousand miles, as the route is now traversed by river and by rail. My own route was in the other direction, from east to west—from Vladivostock to St. Petersburg.

Though, as I have said, the last rail of the trans-Baikal section of the road was laid on December 28, 1899, the Amour and its tributaries were then frozen, and no steam communication could be had until spring unloosed the icy fetters. The rivers did not open this year until the middle of May, and on the 31st day of May I reached Vladivostock, and on the 1st day of June took passage at Khabaroffsk, on the Amour River steamer *Baron Korff*, for the long journey across the Russias.

I mention this fact because the newness of the route will account for some of the crude and primitive accommodations that will be alluded to further on in this story.

To be sure, many other travellers had crossed Siberia from Vladivostock to St. Petersburg, but the journey could not, before May of 1900, be done by uninterrupted steam travel, and thus one more way of going around the world by the aid of the all-compelling steam-engine has been added to the two that already existed.

For many years it has been possible to steam around the Cape of Good Hope and stormy Cape Horn from New York back to New York again. For a generation it has been possible to take the shorter cut by rail across the American continent and by steamer through the Suez Canal. Now a third steam route has been opened which reduces the ocean journey to a minimum, and, for the first time in history, makes the portion of the journey by land considerably longer than the portion by sea.

This is surely good news for bad sailors.

Roughly speaking, this way around the

world involves about thirteen thousand miles of railroad or river travel, and nine thousand miles of sea travel. Approximately, these distances may be divided as follows:

	MILES.
By rail from New York to San Francisco . . .	3000
By ocean from San Francisco to Yokohama . .	5000
By rail from Yokohama to Nagasaki	800
By ocean from Nagasaki to Vladivostock . . .	1000
By rail (and river) from Vladivostock to Liverpool	9000
By ocean from Liverpool to New York	3000

I need dwell but briefly on our journey from New York to Vladivostock, for, so far as Nagasaki at least, it is getting to be as commonplace as a trip across the Atlantic steam ferry. As a matter of fact, it occupied us more than four months, owing to various detours in Japan and China for the sake of attending Christian Endeavor conventions which had called me to the far East. This part of the journey, however, might have been easily accomplished in twenty-five days—five days across the American continent, eighteen days from San Francisco to Yoko-

CATHEDRAL AT VLADIVOSTOCK

hama, and two days from Yokohama to Nagasaki.

Here the novel part of the journey begins. We at once get out of the beaten path of the globe-trotter. He halts on the southern edge of Japan, or, perhaps, more often keeps on eastward across the choppy Yellow Sea to Shanghai, and thence, if he is really a *globe*-trotter, on to Hong Kong, the Straits, Colombo, and Suez.

In the future, however, the stream of world-encirclers is likely to divide at Nagasaki, one part of the stream flowing east and south over the old channel, the other flowing east and north over the new route which, less than three weeks ago as I write these words, was opened up around "this goodly frame the earth."

By this latter route, which I am about to describe, the traveller sails from one glorious harbor to another—from Nagasaki to Fusan, Corea; to Wonsan, Corea; to Vladivostock, Russia. In all of them "the navies of the world might ride," according to the old platitude, and on this coast, without much doubt, will be the naval battle-field of the future, when the Titans get ready to measure their sea-power one against another in the struggle for Corea or China.

We took passage on a little Japanese steamer, the *Ise Maru,* which runs between Kobe and Vladivostock, and which belongs to the Nippon Yushen Kaisha, the largest steamship company in the world, with steamers plying to America and Europe and Shanghai and Hong Kong, besides to the innumerable ports up and down the coast of Japan.

Many of the steamers are large and fine, but the *Ise Maru* is not one of these. She is, however, clean and fairly comfortable, and we will not complain, as the journey is a short one of less than a thousand miles.

A prejudice against Japanese captains and chief officers exists among many foreigners, who will not sail, if they can help it, except on ships manned by European

DRY-DOCK AT VLADIVOSTOCK

officers. I must say that I do not share this prejudice. I have sailed with more than one Japanese captain and crew, and I have invariably found them polite and exceedingly attentive to the comfort of their passengers, and, so far as I could judge, excellent seamen, leaning always to the side of caution in navigating the treacherous rock-bound coasts. As a matter of fact, the steamers that have been lost in these eastern seas have usually been officered by Europeans, and not by Japanese.

Half of the steamers of the Nippon Yushen Kaisha are officered by Japanese, and half by Europeans, and the latter have been by far the most unfortunate.

So, casting prejudice and fear to the winds, and trusting in Providence for a safe voyage, we set sail, and, twenty-four hours out of Nagasaki, approached the bold shores of Corea, no longer the hermit nation, but open now at a dozen doors to the commerce of the world.

Fusan, at the southwestern corner of

the peninsula, is one of these open doors, a door, in fact, which has long been open, for the Japanese for three hundred years have here kept at least a foot in the doorway for their own ingress and egress.

Fusan is in every sense a Japanese port. The Japanese have been here so long that they have even trained the great pine-trees along the streets and on the temple hill to grow in picturesque Japanese fashion, and their shops, their banners, their Shinto shrines, their good roads, all proclaim their long occupation. The only Americans in Fusan are found in a small colony of faithful missionaries, whose homes, on a picturesque hill between the Japanese and Corean cities of Fusan, are conspicuous for miles around. The next day after her arrival the *Ise Maru* weighed anchor again and, shaping her course due north, made for the port of Wonsan, on the Corean coast, midway between Fusan and Vladivostock, and about three hundred miles from either port.

Wonsan, like Fusan, possesses a magnificent harbor, entirely landlocked and surrounded by imposing hills. No wonder that Russia, as well as Japan, has cast covetous eyes upon this splendid port, but to all appearances Japan has the inside track, for here, as in Fusan, she has securely seated herself, has built her houses and shops and wharves and temples, and owns more of the shipping that visits the harbor than all other nations combined.

A few hours only were given to Wonsan, and again the whistle of the *Ise* shrieked, the anchor was hoisted to the bows, and the little ship ploughed her way once more up the bold Corean coast, to make the last loop in her frequently interrupted voyage. One of these days Corea, I believe, will be a favorite haunt of pleasure-seekers and tourists. The coastline is magnificently indented with bays and gulfs, the interior is one series of grand mountains and lovely valleys, the west coast is lined with charming islands that form almost a continuous break to the winds and waves, while the east coast, though without many islands, furnishes equally fine scenery and unrivalled delights for the mountaineer on the shore and the yachtsman on the sea. The people are poor, primitive, and, it must be confessed, dirty, but are kind, courteous, and even courtly, and, in their white linen clothes and queer horse-hair hats, are most picturesque and interesting—far more so than their neighbors in China, *me judice.*

But we are bound for Siberia, and must not in this chapter linger in Corea. Early on the morning of May 29th, five days out from Nagasaki, the passengers of the *Ise Maru* awoke to find themselves in still another magnificent landlocked harbor. One glance at the shore where stands Vladivostock, the "Magic City," showed them that they had left Mongolia and Mongolians behind them, and, though still in Asia, and on the extreme eastern verge of Asia, were in an Asia owned and dominated by Europeans. Large blocks

HARBOR OF VLADIVOSTOCK

of brick and stone crown the hills on which the town is situated. A great white house flying the admiral's flag occupies the most commanding site. A beautiful Greek church, with trim minarets, stands near by. The Russian tricolor is flying everywhere. We had left behind, as completely as though we had sailed to another planet, the mud huts of Corea, the slight one-story wooden houses of Japan, and the stone hovels and filthy streets of China, and suddenly found ourselves in a city of substantial brick and stone, wide paved streets, electric lights and telegraph wires, and railway trains snorting and puffing down to the very wharves. We could almost imagine that either Tacoma or Seattle had bodily taken passage across the Pacific and had planted itself on the eastern shore of Asia instead of the western shore of America. Only the droskies and tarantasses flying through the streets, with one horse under the huge Russian collar while another ran free to keep up with his trotting-mate, recalled us to the fact that we were about to step ashore in monarchical Russia instead of republican America.

A score of sampans, each manned by a long-cued Chinaman, were clamoring to take us ashore, and we were scarcely less anxious to accept their invitation, that we might plant our feet on *terra firma,* and realize that our long ocean voyage was over and our longer land journey over an unbeaten track was about to begin.

GROUP OF KOREAN WORKMEN, VLADIVOSTOCK

II

THE METROPOLIS OF THE NORTH-EAST

VLADIVOSTOCK, on nearer approach, unlike many towns, carries out its more distant promise. It is not a trim and finished town, but it is one of strong outlines, and evidently built for a great future. Its chief buildings are substantial. The town looks as though it had come to stay. To be sure, the new outlet which Russia has obtained through eastern China to the sea at Port Arthur will rob Vladivostock of its unique importance as Russia's only open door on the Pacific, but her glory as the great natural port of Siberia, with a limitless country of vast resources behind her, can never be taken away. The huge ice-breaker which crunches up the harbor ice in winter with its sharp iron beak is a complete success, and it is no longer an ice-bound port for even three months of the year. The winter, though long, is not unusually severe, and the mercury, I am told, seldom or never sinks so low as it does in Boston or New York.

No sooner had we stepped on shore than we were treated to our first disillusionment concerning Russia and the Russians. We had heard much about the terrors of the Russian custom-house. We expected to have our baggage overhauled from turret to foundation-stone, and the feminine member of our party had pictured to herself a swarthy Cossack with sword and cutlass making hay of her feminine belongings, while he searched for dutiable goods. But our Chinese sampan-man landed us at the pier, and no customs officer appeared upon the scene. We waited, but he did not come. We inquired in two or three different languages for the custom-house, but no one knew what we meant. We wandered up and down the pier looking for some building that resembled the familiar office where duties are collected, but could find none. At length, having fully satisfied our consciences, we boldly loaded our *impedimenta* into a drosky, and told the driver to take it to the Hôtel du Pacifique. This he successfully accomplished, and no officer of the law appeared to disturb our peace of mind, or to look into the innocent recesses of our trunks.

So far as I know we were subjected to no surveillance of any sort. We came and went as we pleased. We minded our own business, and every one else minded his. There seemed to be no such inspection of newly arrived strangers as one is often conscious of in Germany. And yet, here we were in autocratic Russia, and, more than that, in Siberia, the land of the exile and the prisoner—the land around whose name has clustered every synonym for oppression and cruelty.

I am inclined to think that what I have been told by an acute observer is very true, and that Russia is much more autocratic and inquisitorial on paper than in reality. Her laws are severe and may be enforced if needs must, but their iron hand is seldom felt by the well-disposed. There is much theoretical restriction, and much practical liberty.

The hotels of Vladivostock are not to be indiscriminately commended. When Baedeker writes his inevitable guide-book about Siberia he will not mark them with his familiar *, unless they greatly improve upon their present condition. It should be said, however, that the best one was recently destroyed by fire, and the chief one that still remains is probably suffering from lack of competition. The table was very good, and the chief fault was that the rooms and halls were dark and dirty, and clean linen was a scarce and precious article. One sheet for each bed seemed to be the full allowance, and one small towel for three people was apparently thought to be an ample supply. When one of our party righteously complained that her towel had evidently done duty before, the Japanese chambermaid, who did the English for the hotel, replied that it was *impossible* to get another; it was not to be had. There was not a clean towel in the hotel. After much importunity, however, she was induced to loan a long, clean scarf from her own bureau to serve as a towel.

However, this was a minor matter. We reminded ourselves that we were in Siberia, and on the eastern edge of Siberia at that. Who could expect all the amenities of civilization? We were no sooner established in our hotel for the day than I sallied out to seek the governor of Vladivostock and of this whole eastern province, to whom I fortunately had a letter from the Russian minister of Japan.

Well was it that I found the governor armed with these credentials, as the sequel will show in another chapter. Governor Tchitchagoff received me with the utmost courtesy. The letter from the minister to Japan acted like an open sesame to his kindly attentions, and he at once despatched a secretary to secure berths on the train which left the next day for Khabaroffsk, and sent another to telegraph to Khabaroffsk to hold a state-room on the Amour River steamer for Blagovestchensk, as he knew the berths were in great demand. At the same time he gave me a letter to the chief of police in Khabaroffsk. While waiting for the preparation of this letter by his secretary he showed me a map of the vast province of Primorskaia, of which he is governor, a province which extends for two thousand miles up the shores of the Pacific, clear away to the North Pole, embracing Kamchatka and the famous Saghalien Island north of Japan, to which are banished the worst of Russian convicts.

Governor Tchitchagoff is a man of commanding presence, but most urbane and gentlemanly, a trait which he shares with most Russians of the better class, so far as my observation goes. There is a cer-

GENERAL TCHITCHAGOFF

Governor of Vladivostock and Primorskaia

tain courtly simplicity about them, free from all arrogance and brusqueness, which is the very essence of high breeding. The governor speaks fluently French and German, and English with a little hesitation. This, again, is a type of the educated Russian of the better class, who is usually a famous linguist.

Moreover, the governor is a Cossack, a name which, in my ignorance, I had hitherto associated largely with frontier troopers of the Buffalo Bill order. I confess my stupid misconception with contrition of spirit. Thus again came a happy disillusion concerning Russia and the Russians. This interview with the governor set my mind at rest in regard to any danger of losing the steamer on the first stage of our long river journey, for I knew that the word of General Tchitchagoff was as the law of the Medes and Persians, which altereth not. He is the virtual czar of all this Eastern land. It was exceedingly necessary that we should incur no long delays, as a most important engagement awaited me in London within forty-four days. I had already cabled "London screaming," which meant, according to the cable code, that I would be in London on the Fourth of July (a most appropriate cable word for the date in question), and there were very few days to spare over the shortest time on record. Mail steamers on the Amour River start on their various stages only once in five days, and it would not do to miss connections at the outset.

But now, with easy minds, we set out to "do" the town. It does not take long to exhaust its glories, for it is a new town, in the raw, half-baked stage of its existence, with most of its glories in the future. But these are evident enough to make up for any present deficiencies. Vladivostock is credited, in the latest published statistics, with about 29,000 inhabitants, but it is now said to have 40,000, chiefly Russians, Chinese, and Coreans, with a sprinkling of Japanese, Germans, French, Italians, English, and a very few

Americans. It is a remarkably cosmopolitan little city to be found on the banks of an almost arctic sea, and it exhibits all the life and picturesqueness of a little cosmopolis. Russian droskies go flying about the streets in every direction, as though their occupants were pressed for the last second of time, and would be eternally too late unless they arrived somewhere within less than five minutes. Most harum-scarum, reckless drivers apparently are the drosky men. Their horses are given a loose rein, and they go flying down-hill at breakneck speed, over the rough roads, and galloping up the next incline with never a check on their mad career. The appearance of reckless haste is further increased by the fact that always one horse and often two horses are hitched to the vehicle outside the shafts, by loose ropes or leather thongs. These horses often pull away from the wheel-horse within the shafts so far as the harness will let them, and they dance and prance and gallop while he trots at full speed, and the drosky goes rattlety-bang over the stones and into the holes and over the "thank-ye-marms" in a way that the Russians seem to like, but which tires the patience and, let us hope, improves the liver of the foreign passengers.

Besides the drosky, the streets are full of another peculiarly Russian vehicle, the isvodschik, which resembles a small boat set on long poles, somewhat after the fashion of the body of a buckboard. This is used to carry luggage, and, in many parts of the country, passengers as well, who pile into it until it looks like a peasant's cart in Naples or a Chinese wheelbarrow in Shanghai. The Chinese wheelbarrow is not wanting in Vladivostock either, and its familiar squeak is heard on every hand. Vladivostock, indeed, must be the paradise of the Celestial, for here he gets a wage of a rouble and a half a day, which is something like ten times the pay he would receive in his native land. Moreover, there seems to be unlimit-

GENERAL VIEW OF VLADIVOSTOCK

ed work for him in grading and paving the streets, carrying brick and mortar, rowing the sampans, building the railways, and doing the other drudgery of a new colony. The Corean, in his dirty white garments and padded pillow-cushion shoes, is not far behind him in numbers, and competes for the manual work of this new country with his Chinese cousin on the same terms.

Vladivostock consists of one chief street, running up-hill and down-dale, with various branches climbing the hills on every side. The chief sights are easily done, as I have said, for they consist of a few fine stores and public buildings, a beautiful Greek church with its graceful domes, and a triumphal arch in honor of the visit of the czarewitch, a few years ago, when the first spadeful of earth, in breaking the eastern end of the new Siberian railway, was turned by the royal visitor.

But Vladivostock's chief glories do not lie in her public or private buildings, or in the great fortresses which command the city and harbor, but in the magnificent scenery that greets the eye on every hand. Few cities in all the world are more beautiful for situation. The Amour Bay, on one side, and the Bay of Ussuri, on the other, clasp the city in their kind embrace, and constantly lave its feet with gentle ripples. Noble hills stretch off landwards, and from the main hill-tops within the city limits one gets glimpses of sea and shore and mountain and valley of surpassing loveliness. In view of the long journey to which we were looking forward, the railway station, prosaic as it is, was the most interesting building in all Vladivostock. Over its doorway is printed, in old Russian characters, a legend which may well stir the blood as we think of all that it means. It reads as follows:

VLADIVOSTOCK TO PETERSBURG
9877 VERSTS.

We were to begin to measure off these versts on the morrow.

III

RIDING ON THE RAIL IN SIBERIA

OUR first introduction to a Siberian railway train was made under unusually pleasant circumstances. Hurrying to the station in ample season to secure what accommodations we could, and expecting to have a severe struggle with our tickets and our baggage in a strange language of which we knew scarcely a single word, what was our surprise to have a decorated official, in a handsome white uniform bound with gold lace, accost us by name and assure us that, at the governor's orders, a first-class compartment had been reserved for us, and all we had to do was to buy our tickets and take our seats. As the train was very much crowded, and our lack of knowledge of the language made us almost as helpless as infants in looking out for our comfort and advantage, this was a favor which we most thoroughly appreciated. Before the train started we were favored with a call in our little state-room from the enterprising and energetic United States Consul, Richard T. Greener, and also from Mrs. Frederick Pray, a little American lady, who, a few years ago, transferred her home half around the world, from Berwick, Maine, to Vladivostock, Russia. Her husband is connected with a large American firm doing business in Vladivostock, that is introducing many products of American soil and a thousand Yankee conveniences to the Siberian public. Mrs. Pray had kindly prepared for us a list of common Russian phrases, with the English pronunciation, which proved to be exceedingly useful in calling for tea and bread-and-butter, and a carriage, and a room at a hotel, or any one of the necessities of life to the wanderer in Siberia, where French is scarce, and German scarcer, and English almost nil.

STATION OF KORFOVSKIA, EASTERN SIBERIA, SHOWING AMERICAN LOCOMOTIVE

The Russian language is formidable enough even when one has time and opportunity to make a study of it; to the passing traveller it is absolutely appalling. It has thirty-six letters instead of twenty-six, and several of them seem absolutely superfluous, for they cannot be and never are pronounced. The Greek scholar at first hails the alphabet with delight, and tackles it with enthusiasm, for he recognizes several old friends. There is delta and lambda and phi and chi and theta and several other letters that he struggled with in his school-days. But, alas! his enthusiasm is soon quenched, for he finds that there are so many other letters that are neither Greek nor Roman that he is more confused than helped by the presence of his old friends in such strange company.

Moreover, several of the letters are exactly like Roman characters, but have a very different sound, which is very confusing. For instance, B is not B at all, but is V, as in vein, while the Russian B is an entirely different character. H is not our familiar aspirate, which Cockneys so sorely abuse, but is our N. P is equivalent to double R in hu*rr*y, while the Russian P is made like the Greek pi. C is always S, and there is no hard C in Russian. Y looks very familiar, certainly, but is really oo, as in m*oo*n, and a letter that looks like R wrong side before, **Я**, is the vowel ya, as in *ya*rd. Then there are certain italicized letters which still further bewilder the puzzled student, for u = i and n = p and m = t when written in italics, but not otherwise. Thus it will be seen that Russian, "as she is spoke" or as she is wrote, is not by any means a holiday affair, and the usefulness of Mrs. Pray's extemporized phrase-book can be fully appreciated.

But the time has come for the start, of which we have been made fully aware by numerous bells and gongs and whistles. There is no excuse for any one to get left at a Siberian railway station. Five minutes before the train starts a large station bell is rung. Four minutes more the passengers stroll up and down the platform or visit the buffet. Then the bell is rung once more, the conductor blows his whistle, the engine shrieks a warning blast, and at last we are off, with St. Petersburg 9877 versts (more than 6250 miles) away.

The scenery for a few miles out of Vladivostock is superb. The railway skirts one of the great bays between which the city lies, high up above the water, while off in the distance, beyond the blue ripples, lie the blue mountains, only partially obscured by the morning haze. Chinese junks, loaded with firewood for the locomotives, are dancing on the waves, or unloading their cargo by throwing the pieces into the sea for the waves to wash ashore, since the shallow water prevents the junks from coming close to the bank. Occasionally the railway diverges from the shore and runs through a copse of birch or beech, fresh and bright in their new spring livery, and then returns once more to give us a glimpse of the bright blue sea. The day was charming. Our most delicious spring weather in America is no more delightful than this first day of June in frozen Siberia. Had James Russell Lowell lived in Siberia, he would have written with equal rapture his oft-quoted lines:

"What is so rare as a day in June?
Then, if ever, come perfect days."

On one of these "perfect days" it was our good fortune to journey from Vladivostock. The spring seemed just about as far advanced as it would be in northern New England on the same date. The meadows were lush and rank in their growth, and the cattle waded knee-deep in their delicious fodder, and were, of course, as sleek and fat as cattle could be. The trees were nearly in full leaf, though some of the later varieties had not as yet donned their full suit of green. Great dandelions, almost as large as peonies, starred the fields with yellow, and

"WHO'LL BUY MY FLOWERS?"

Lilies-of-the-valley at the railway station

bluebells and tiger-lilies made the roadway gay.

At the stations little girls with bare feet offered great bunches of lilies of the valley for sale, and I noticed that they were not unlike the rest of the world in taking advantage of the unsophisticated traveller, for, while I paid fifteen kopeks for a bunch, a Russian by my side, who knew how to dicker and bargain in the vernacular, paid but three kopeks for his lilies. But every traveller in a strange land must pay dear for his experience as well as for his lilies. Let me say, once for all, that one hundred kopeks make a rouble, and that a rouble, at the present rate of exchange, is worth a little more than fifty cents. Russian money is on the gold basis, and is in gold, silver, paper, and copper. National bills up to the value of five hundred roubles are good in any part of the empire, and are in denominations of three, five, ten, twenty-five, fifty, one hundred, and five hundred roubles. Gold is in common use in pieces of five, seven and one-half, ten, and fifteen roubles, and silver in roubles, half-roubles, and twenty-five, twenty, fifteen, ten, and five kopek pieces, while copper coins are of one, two, three, and five kopeks in value. In my opinion, there are too many coins of different values, and one and five kopek pieces in copper, and ten, twenty-five, and fifty kopek pieces in silver, and five and ten rouble pieces in gold would answer quite as well the purposes of trade as the larger variety now in use. As it is, the fifteen, twenty, and twenty-five kopeks are so nearly of the same size as to be confusing, especially as different issues of the same value are not always of exactly the same size. The older silver coins seem to be of debased metal, with a large admixture of copper, but the newer coins that bear the image and superscription of the present czar are of much better metal.

After the edge of the novelty of our first ride on a Siberian railway wore off we had time to examine the railway equip-

VILLAGE OF VIAZEMSKOI, EASTERN SIBERIA

ment and the carriages in which we were riding. The train was drawn by a sturdy Baldwin locomotive, fitted with a big bulbous smoke-stack for burning wood, and consisted of about a dozen cars of the first, second, and third classes, in which the third-class largely predominated. The third-class were painted green, the second-class yellow, and the first-class a dark shade of blue. Sometimes a single car was divided into two classes, first and second, one-half being painted blue and the other yellow, with very little difference between the two classes except the color of the paint and the price of the tickets. The second-class cars were, perhaps, upholstered a shade less luxuriously than the first, and were somewhat more crowded, though there was little to choose in this respect. The first-class fare was seventeen roubles (about eight dollars and a half), and the second-class about one-half that sum. We were told in advance that the trains were built on the "American plan," but it is really a combination of the American and English, somewhat in vogue on the continent of Europe. All the first and second class cars are divided into compartments, but down their side runs a narrow corridor wide enough for only one person to squeeze through at a time. The compartments open on this corridor, and through it the conductor passes to collect his fares; passengers walk back and forth, babies escaped from their nurses toddle down, and it is the general thoroughfare for a heavy-laden train. Being exceedingly narrow, it is always "congested," and a fat man finds it his "misery," indeed. In order to get through at all he must turn sideways and go crab fashion. There seems to be a great social democracy in this aristocratic country, and passengers from the different classes visit back and forth with the utmost freedom, and occupy compartments of a higher class than their own for half a day at a time.

As a rule, the Russians are exceedingly polite and considerate of other people, and,

THROUGH THE FOREST IN EASTERN SIBERIA

for the most part, the rude and selfish people whom we have met on this journey have belonged to other nationalities. The third-class cars frequently have an aisle down the middle, with seats on either side. As all the coaches are meant for night travel, all provide what is called "lying-down room"—*i.e.*, room is provided for each passenger to stretch out at full length either in the upper or lower berth. The first and second class cars are luxuriously upholstered, and, by a curious contrivance, the upper bed turns over and becomes, not "a chest of drawers," but the back of the seat by day. The only real lack of a Siberian railway is suitable lavatory accommodations. The little toilet-room is often a wretched, filthy closet, with a single wash-basin and a very limited supply of water, and it answers for all, men and women alike. This fault is shared by all Siberian hotels and steamboats that I have seen. The one cramped and dirty spot is the wash-room (for many hotels have a common lavatory, and no water is brought to the rooms), and the one scarce article is fresh, cold water. Beer, wine, vodka, tea, especially tea, flows freely, but to order a glass of water to drink, or a basin of water, much more a tub of water for a bath, creates a commotion, and the water desired is often unattainable, except after strenuous effort. A Siberian writer remarks, naïvely, that "Englishmen have the bad habit of washing themselves all over every day. *As a consequence of this habit,* their bodies emit an unpleasant odor." It must be said in all fairness, however, that hotels, cars, and steamers in Siberia are, outside of the wash-rooms, clean and wholesome for the most part. To one who has just come from China or Corea, and who has tasted the rare discomforts of travel in those countries, they seem to be beyond criticism.

Besides the cars already mentioned, a baggage-car and a dining-car completed our train equipment. The latter, at least three times a day, is a place of much interest, and deserves a few words of description.

Pullman would scarcely own the diner as an offspring of his invention. A long table down the middle, at which perhaps twenty people can sit at one time, and a bar at the end, at which all kinds of light and strong drinks are served, and toothsome delicacies dear to the Russian heart, like caviare, sardines, and other little fishes "biled in ile," are eaten. At the long table *table d'hôte* meals are served, consisting of three or four courses, and one can also order what he chooses, at a fixed price. The meals are usually one rouble each, and though not luxurious, from the American dining-car point of view, they are quite sufficient for the average traveller, and worth the price asked for them. To be sure, one must get used to the greasy Siberian soup and to the chunks of tough stewed meat, which may be beef, mutton, or pork, one is never certain which. But travellers who choose to go across Siberia should not be squeamish, and to think of eating in a dining-car, however primitive, while whirling across the plains a little south of Kamchatka, is enough to kill the spirit of criticism in the most confirmed growler who ever went around the world.

It only remains to be said that the road-bed of the Eastern Siberian railway is well made, but seems to lack permanent stone ballast, the bridges are substantial iron structures, the piers and culverts are thoroughly built of cut stone, the embankments are smooth and well sodded, but the rails are light, and, it would seem, must be replaced by heavier ones before any high rate of speed can be attained.

CITY OF KHABAROFFSK, ON THE AMOUR RIVER

IV

FROM VLADIVOSTOCK TO KHABAROFFSK

THE distance from Vladivostock to Khabaroffsk is a little less than five hundred miles, and it requires fully twenty-nine hours to cover it, so it will be seen that the trains are not run on the "Empire express" plan. However, before we had been many days in Sibèria such a rate of progress had come to seem speed itself. The stations on the eastern end of the Siberian railway are not many, for the villages are few and far between, but we make up for the fewness of the stops by spending a long time at each station. Still we can occupy the waiting time very pleasantly by getting many, though fleeting, glimpses of Siberian life. The stations themselves are usually wooden buildings, neat and attractive and prettily painted. A picturesque water-tower is always a feature of the grounds, and usually an underground store-house, sodded over, that looks like an Oklahoma cyclone cellar. In this, doubtless, provisions are kept from freezing during the long Siberian winter. All the principal stations have buffets, where thirsty Russians refresh themselves with vodka, kvass, and other national drinks.

The villages usually consist of a few log-houses, with an occasional more pretentious frame house, but they are neat and attractive, and compare very favorably with the villages on our own Western frontier. The most interesting object at all these stations was the line of country men, women, and children who had brought the products of their farms to sell to the hungry travellers. Every station had its hucksters, not noisy or obtrusive, but waiting patiently in an orderly line for customers. Bottles of milk for ten kopeks, hard-boiled eggs for ten kopeks a dozen, huge loaves of black, sour bread in which the Siberian rejoices, chunks of fried meat and fried fish, huge pickled cucumbers, bottles of home-brewed kvass, a kind of innocent pop beer, were among the comestibles offered for sale. The hucksters were all hearty, healthy specimens of humanity, not refined or elegant in feature or dress—this could scarcely be expected in this new land—but wholesome, honest-faced settlers, capable, evidently, of laying deep and strong the foundations of an empire. I am told that many of them are dissenters from the Greek Church, and that they live godly and simple lives, and are much respected by their neighbors, and in this part of the country at least are unmolested by the government, unless they try to proselyte among adherents of the national faith. Most of them, too, are vegetarians. Every village has its church, often embowered in a little park of green, and the towers of blue and green, colors which the Siberians greatly affect for their churches, and bright copper domes, shining in the brilliant sun, always gave a touch of pleasing color to the scene, and showed that the people were not wholly given to buying and selling and getting gain in these little raw railway settlements.

The costumes of the people who flocked to the railway stations were most picturesque and interesting. Red was decidedly the favorite color with the women, and red of every shade and hue. Dull red and bright red, Turkey red and blood red, crimson, magenta, scarlet, and crushed strawberry. Often several, if not all, of these colors were combined in the same costume in a way which, I suppose, would have set the teeth of our fashion-plate ladies on edge, but which, after all, was undeniably picturesque.

The men were usually clad in a loose blouse, belted around the waist, and loose trousers tucked invariably into huge top-boots that reached nearly to the knee. Many times these boots looked to be by

FATHER ALEXANDER AND STATION-MASTER AT NIKOLSKOIE

far the most expensive feature of the costume. As a Mexican peasant will often wear a sombrero, it is said, worth as much as his coat and trousers and boots, mule and saddle—in fact, the whole outfit combined—so the Siberian peasant clothes the other end of his anatomy with equal care, and even if he must go in rags, will bankrupt himself on his boots.

At all these stations the military were strongly in evidence, as is not unnatural, since this is a military road and the large places are all military headquarters. Everywhere, in the cars and out, officers and soldiers were to be seen. At least every other man you met seemed to be in uniform. But they, too, were invariably polite and showed nothing of the haughty arrogance and devil-may-care air often exhibited by the soldiery of Germany and other army-ridden countries. In fact, I was struck on this railway ride, and on all my other journeys in Siberia, by a certain quiet, dignified, self-contained air that the people displayed. The hucksters are not insistent on your buying their wares. The cabmen are not vociferous. The few beggars are not obtrusive. It is as though the vast, lonely stretches of their homeland had somehow impressed themselves upon the character of the people. The broad reach of upland and prairie, the noble hills, the wide stretch of blue heavens above them, the long, serious winter, the strenuous conditions of their life in this virgin land, have developed a national character, which seems never to have acquired the noise and bustle and artificial "hustle" of city-dwellers. On the whole, it is a pleasing, restful, and, I should think, likable type of character to which these surface indications point. I felt that I was back in the early days of New England, among the Pilgrim fathers and mothers of a new empire, as I looked into these earnest, frank, and serious faces, and as I looked I had large hope for the empire they are founding.

The parish priest is also a picturesque and interesting factor of village life. He often came to the station to take part in the one excitement of the day, the arrival of the daily passenger train. He was usually clad in flowing robes of black, with a close-fitting purple cap, while a huge gold or silver crucifix dangled from his neck. His hair is always long and curly. Whether the Greek priests were chosen for their curly hair or indulged in the use of curling-irons when preparing their morning toilet was a question which interested the ladies in our party. They seemed to mingle with their people on terms of benevolent equality, and their kindly and gracious presence shed a benediction even upon the railway platform. They were not averse to having their pictures taken, and the camera enthusiasts "snapped" the holy fathers with the greatest freedom.

The kodak is not unknown in Siberia, but it is evidently much more of a novelty than in the west, and always attracted crowds of staid peasants who wanted to look into the "finder" and see the resultant picture. Some were quite disappointed that the completed picture could not immediately be taken out of the camera and shown to them. One gorgeous railway official, much be-braided and be-spangled with gilt, insisted on having his black-bearded face transferred to the sensitive film. He posed with his most radiant expression, without any exhortation to "look pleasant." He followed our party about, and apparently expressed every good wish for our successful journey, though our knowledge of Russian was not equal to his voluble good wishes, and, finally, just as the train pulled out of his station, he handed us a paper on which he had written a request that we telegraph him of our safe arrival at Blagavestchensk, a city on the Amour River, a thousand miles further west. A kindly fellow-feeling that, to be provoked while the train stopped "ten minutes for refreshments."

Fifty miles after leaving Vladivostock the railway climbs to a considerable pla-

A PATHETIC SIBERIAN VILLAGE GRAVEYARD

teau, and the trees disappear, while open prairie land takes the place of the forests that crown and gird the hill-sides. For a couple of hundred miles, at a rough guess, the open prairie land continues. The soil mostly seemed to be a rich black loam, and one would think it was capable of supporting millions of people where it now supports hundreds.

On the second morning of our journey, twenty hours after leaving Vladivostock, we woke up in a heavily wooded, hilly region, and I had to rub my eyes to make sure that I was not dreaming and had really awaked in Siberia and not in northern Maine. Here was the Moosehead Lake region duplicated, apparently, in every stump and white birch-tree and corduroy road. Here were the same tree-clad hills and rocky ravines, with glimpses of other and higher hills peering above the tree-tops. Here were the same clear running brooks, babbling over their rocky beds, the same patches of charred timber burned to clear the land, the same flora down to the dandelions, blue-bells, buttercups, and cowslips. To be sure, there was not such a predominance of spruce-trees as in northern Maine, but their places were taken by the larch and cedar and other evergreens, and white and yellow birch, beech, and poplar, which predominated, made me feel very much at home.

For hours these familiar forests lined the railway, and then the train emerged into the fertile plains that border the Amour. Here the snow fences, piled up in their summer quarters, reminded one of the prairies of our own Northwest, and showed that in the Siberian winter the railway has the same great enemy to fight. The telegraph poles, often braced by two supports until they look like so many tripods, also testify to the strength of Siberian winds. At length, at noon of the second day, twenty-nine hours after leaving Vladivostock, the spires and chimneys of Khabaroffsk appear across the prairie, and we realize that the first stage,

THE OLD WAY BY THE "SHIP OF THE DESERT"

and a very pleasant one, of our trans-Siberian journey was nearly finished.

Scarcely had the train pulled into the station of Khabaroffsk, and before I could get to the door of the car, when a police official in gorgeous raiment crowded in and inquired anxiously for "Pastor Clark from America." When I confessed to being, in all probability, the individual that he sought, the official seemed much relieved, and at once took charge of me and mine, bag and baggage, and summoned a small squad of soldiers, or sub-police, to assist him. Evidently the kindly telegram of Governor Tchitchagoff was doing its beneficent work, and, like expectant American politicians, we were "in the hands of our friends."

V

THE *BARON KORFF*

OUR friend in need proved a friend indeed, for it would have been no easy thing, knowing scarcely a word of the language, to disentangle our baggage, secure a cart to carry it, and droskies to carry ourselves, and make our way through the city of Khabaroffsk to the steamboat landing two miles away in season to catch the mail steamer, which, with steam up and whistle shrieking, was waiting to start on her long journey up the mighty Amour. But our official accomplished it all for us with celerity and despatch. He picked out our baggage, hired our droskies, took another himself, and drove to the steamer, saw our baggage safely bestowed in a state-room which had been secured for us at the gov-

THE *BARON KORFF*

ernor's orders, and did not leave us until we were fairly established in our comfortable and commodious quarters on the *Baron Korff.* We appreciated this courtesy the more fully when we learned that the steamer was exceedingly crowded, that it went only once in five days, and that some people had already been waiting fifteen days for the coveted passage. Our rapid flight through Khabaroffsk was made in a cloud of dust which would have done honor to Peking itself, the dustiest city in the world. Through the dust, however, we were able to catch glimpses of a frontier American-like town, with some good business blocks of brick and a fine cathedral church. We afterwards learned that Khabaroffsk contains about fifteen thousand inhabitants, and its importance is derived from the fact that it is the highest seat of government for all this vast section of the Amour River. Sitting on a high bluff at the juncture of the Amour and the Ussuri, a great stream which here joins its forces with its greater rival, the town occupies a commanding position and is regally situated to dominate the traffic of both rivers. In any other land the Ussuri itself would be a notable river, but in this country of great rivers, though it is nearly a mile wide at its juncture with the Amour, it is only a branch of eastern Siberia's greater waterway.

Since the *Baron Korff* was to be our home for the next week, and, in a literal as well as figurative sense, we were to be "all in the same boat" with the rest of her passengers, we were naturally anxious to know something of them and of her, if it is proper to speak of a baron as belonging to the feminine sex. A hasty inspection showed us that the *Baron Korff* was a stately side-wheeler, such as might have done duty on the Mississippi before the railroads robbed her of her commercial glory. The first-class accommodations were well forward, and consisted of roomy and comfortably furnished staterooms for about twenty people on the upper deck, each room having berths for two people, a long dining-saloon, into which the state-rooms opened, and a pleasant saloon in front, handsomely furnished and provided with a good German piano. This saloon made a delightful sitting-room for the passengers, for its abundant glass furnished a splendid view of the river and the banks on either side, between which the *Baron* made her devious way. Behind the first-class cabins was a large open space, sheltered from the wind, and suitable for a promenade. Here was a restaurant and bar, frequently patronized by some of our thirsty fellow-passengers, while behind it were the second-class cabins, which looked to me nearly as comfortable as those at the other end of the ship. Down below were huddled together a miscellaneous crowd of peasants—men, women, and children—with their scant piles of bedding and all their worldly possessions; sturdy men in top-boots and figured gowns belted at the waist, often with a sleeveless vest worn outside; women in gay calicoes and hobnailed shoes; children clad in little more than nature's original raiment—all were here. On the hard iron floor they spread out their quilts, piled their household goods about them, and prepared to make the best of their six days in the bowels of the *Baron Korff*.

Their accommodations were not luxurious, certainly, but when we remember that they paid but three roubles and six kopeks for the entire 918 versts, more than six hundred miles, to Blagavestchensk, or only about one dollar and a half, we cannot think that the accommodations were disproportionate to the price.

The chief rival of our friends in the steerage was the great pile of wood, for which the *Baron Korff* had an insatiate maw, and which, when first put on board, occupied fully one-half the space of the steerage, but which, every hour, sensibly diminished until we came to a station, where a new supply was taken on board.

The second-class fare was four times the third-class, or thirteen roubles and

EMBARKING ON AN AMOUR RIVER STEAMER

ninety-seven kopeks, and the first-class more than seven times the third, or twenty-two roubles and ninety-six kopeks. I am constrained to admit, however, that the first-class accommodations are fully seven times as good as the third, though I do not think they are twice as good as the second.

For the meals on the Amour steamers one pays separately, at the rate of two roubles a day, not an extravagant sum, indeed; but they were not extravagant meals, and they cost all they were worth.

In the morning, about eight, tea was served, with the inevitable samovar occupying the place of honor in the centre. Slices of lemon and sugar were provided, according to the Russian style, but no milk. Soft bread and zweiback, without butter, completed the first meal of the day.

The second, and principal, meal was served promptly at noon, and consisted of a thick, greasy soup, a course of meat with no potatoes, and very scanty vegetables of any kind, and a pudding. On rare occasions ice-cream took the place of the pudding. Tea was served again at four o'clock, and supper, which consisted of one meat course without vegetables, and more tea, completed the menus for the day. If one wished to supplement these meals with orders from the restaurant of caviare or smoked fish or boiled eggs, of course he was at liberty to do so.

The first meal served on the *Baron Korff* on our voyage was supper. Fifteen foreigners and five Russians sat down to the table together. None of the foreigners, at least, knew what to expect from the culinary department. The meat was served, and was sparingly partaken of, on the supposition that, after the fashion of other steamers, four or five other courses would follow. Then came a long wait. The plates were removed, and still we waited. The minutes lengthened into a full quarter of an hour, and at length the one overworked waiter appeared, bearing, not the expected fish or vegetables or salad or dessert, but *the brass samovar,* and we knew a tumbler of tea was all that was to be expected. The craning of necks as the long-delayed waiter entered the room, and the expressions of disappointment when his sole burden was discovered, were amusing. But though some went hungry to bed the first night, Experience, the mother of wisdom, taught a needed lesson for all travellers on the Amour River—to lay in a sufficient store from the first platter, and not trust to the "hereafter." After that no one omitted the first course.

I dwell somewhat at length on these details, because, so far as my experience of Siberian river-steamers goes—and this experience became somewhat large and varied before we reached Stretinsk, twenty days later—they are characteristic of all river travel in Siberia.

The steward, to whom I have already alluded, was a sadly overworked individual. He was waiter, cook, bell-boy, chambermaid, boots, and man-of-all-work for all the twenty first-class passengers, and, as can be imagined, no grass was allowed to grow under his feet. To be sure, his duties as chambermaid were merely nominal, for every one took care of his own room, furnished his own bedding, and made his own bed, and all had to wash in a common lavatory and furnish their own towel, soap, and, of course, all toilet articles. This common washroom arrangement was the most unpleasant feature of the *Baron Korff*, and we had occasionally to remind ourselves that we were in the extreme boundaries of Siberia. Who, then, had a right to look for all the luxuries of the latest civilization?

At length we were embarked on the long river journey to which we had so long been looking forward, and about which we had heard such various and contradictory reports. For months we had made it a point to inquire of every one likely to know concerning the trans-Siberian route. No two

ON THE SHORE OF THE AMOUR

people agreed concerning it, or came within sight of agreement. "It can't be done," "You can do it easily," "It will take two months," "You can go through in twenty-two days," "You will get stuck on the sand-bars for weeks," "You will have no difficulty whatsoever," "The steamers run only occasionally, and do not begin until the 11th of June," "They run every day, and the river is open early in May," "You will need heavy clothes, and all your winter furs," "You will find delightful summer weather," "You will have to ride in cattle-cars when you get to the end of your boat journey," "You will have the most luxurious railway accommodations in the world."

Such were the varied and miscellaneous bits of misinformation which we had been able to pick up in America, Japan, China, and even in Vladivostock itself. But now the journey on the great waterway of Siberia had begun. The paddle-wheels of the *Baron Korff* had begun to revolve, we were to be among the first few score of people who had crossed the Russias by steam, and we should soon know which of these conflicting stories were correct, or whether they were all alike untrustworthy.

VI

THE LORDLY AMOUR

FAIR as was the scene that first bright June day on the Amour, we could not altogether forget that we were in Siberia, for, as the lines of the *Baron Korff* were cast off, another steamer of like pattern anchored just beyond moved out of our way, and we saw that she was towing a convict barge, and the faces of the poor fellows bound for dismal exile could be seen, pressing close up to the bars of their floating cage. These prisoners were bound for Saghalien Island, a thousand miles away, at the mouth of the Amour.

No sooner had the *Baron Korff* gotten clear of the pier than we found that she, too, towed a convict barge, and henceforth, for days and days, for hundreds of miles, one of these grewsome ships followed in our wake, tugged by the same steam-power that carried us along. But what a difference! We were journeying in comparative luxury, with pleasant company, and all the freedom that a roomy steamer can furnish, to friends in Europe, and then to our home in the home land.

They, shut up in a white, floating cage, in narrow quarters, were facing exile and misery, and perhaps death, far from their homes and kindred. Every revolution of the paddle-wheels was carrying us nearer home, and them farther away. What a mockery the bright sunshine and green, tree-clad banks, and boundless, free prairies, stretching beyond, must have seemed to these cabined, cribbed, confined, hemmed-in voyagers!

It must be confessed that these particular convict barges from a distance did not look either filthy or overcrowded, and closer inspection afterwards convinced me that if prisoners must be exiled, they could scarcely have expected more com-

A PRISON BARGE

fortable conveyance. Still the thought of those fellow-travellers, so near and yet so far from the view-point of freedom, hope, and good cheer, cast a shadow on the journey whenever we looked at their white prison trailing on behind.

The Amour at Khabaroffsk, though we were more than five hundred miles from its mouth, was fully a mile and a half wide, and flowed in a strong, full current, which fact we realized for many a day thereafter, as we made our slow and toilsome way against it. The Amour is one of the few greatest rivers of the world. In length it is equalled by no river in Europe, and surpassed only by the Yang-tse-Kiang and Yenisei, in Asia, by the Nile and Congo, in Africa, and by the Amazon and Mackenzie, in America, though, if we reckon the Mississippi and Missouri as one river, it is longer than any of them except the Nile.

The basin drained by the Amour contains over 500,000 square miles. It is a singular fact that so many of the great rivers of the world are of nearly the same length. The Amour, Yenisei, Mekong, Obi, Hoangho, and Lena, in Asia, are all within two or three hundred miles of the same length, and all surpass 2500 miles.

The Congo, according to the geographers, is only about a hundred miles longer than the Amour, while the Niger is two hundred miles shorter. The Amazon is three hundred miles longer than the Amour, the Mackenzie a hundred miles longer, while the St. Lawrence falls short of it by six hundred miles.

But, in a general way, it may be said that a dozen of the great rivers of the world are practically of the same length. Of them all, few are more majestic and imperious in their flow, or drain a fairer country than the Amour.

Its water is somewhat muddy at Vladivostock, but nothing like the consistency of the Mississippi at St. Louis. It cannot be said to be " both food and drink." As we ascend its swift current it constantly grows clearer, until, a thousand miles farther up, it is about the color of white wine, and is sweet and wholesome to the taste.

The Amour is unique in the fact that it flows through a fertile country which is almost uninhabited, and yet is capable of supporting millions. No great cities line its banks, and not many towns that rank larger than hamlets. Even these are few and far between, and one journeys for half a day at a time without seeing even one little hut among the bushes. Its shores are still in their virgin greenery. For hundreds of miles at a time one sees not a cultivated field, though doubtless some such tilled land lies back from the river and out of sight. Great wood-piles for the use of the frequent river-steamers are the most common objects that show the hand of man, but even these are often in desperately lonely spots, so that one can scarcely believe that they were ever visited by human beings. And yet all this immense river basin of half a million square miles is apparently fertile and habitable, and, when we saw it, was glowing in rich and brilliant verdure. Surely the world is not yet over-peopled while such a lordly domain is waiting for the plough and the reaper.

To be sure, we saw the Amour Valley in early June, when everything in nature is at its freshest and best, and our judgment might have been modified had we seen it in December. Still, every country, like every man, should be judged at its best and not at its worst.

Promptly at four o'clock the *Baron Korff* started on her long journey, and went zigzagging across the lordly river in order to stem the swift current as best she might. On the left-hand side was Chinese Manchuria, on the right Russian Siberia. But the Bear has her huge paw on all this country, and only by sufferance has she for many years allowed the Sick Man of Asia to nominally rule the left bank of the river.

It is a suggestive journey that we are taking on the broad ribbon of water that

A SIBERIAN RIVER TOWN IN WINTER

divides two of the greatest nations of the world—the one hoary and decrepit, with the weakness of her centuries heavy upon her, the other lusty and stout in comparative youth, like a strong young man lately come to the consciousness of virile powers.

Two hundred years ago and more China and Russia tried conclusions on these same plains of the Amour, and then China came off best. Russian garrisons were captured and carried off to Peking, and a treaty was concluded ceding all this land back to China. But during these two centuries China has daily been growing weaker and more corrupt in her senility. Russia has become mightier with each succeeding year.

In 1868, by the treaty of Aigun, so called because of the Chinese town on the banks of the Amour, where the treaty was signed, China ceded all the country north of the river to the Great Bear, who afterwards obtained concession after concession, until she was paramount in all Asia north of and including Peking.

Then came the last uprising of China, which began the very month this journey was made, and which again threatened Russia's domination, at least for a time. Thus the shuttlecock of dominion over northern Asia has been thrown back and forth between the battledore of Russia and China, and the end is not yet.

It will be seen that if this country through which we are making our zigzag route is lonely, it is not devoid of historic interest. It has been and is a battle-ground for armies and diplomats, and is likely to play a still larger part on the world's checker-board in the future.

All through the first afternoon and through the long twilight which, in these northern latitudes, leaves but a few hours to the night, the *Baron Korff* ploughed her way along the mighty river which seemed to be vexed by no other paddles. But so great is its extent that, although there are hundreds of steamboats on the river, to meet one is still an event.

During the first part of the voyage the

Manchurian shore presents the most striking scenery. Many bold and picturesque hills come down to the water's edge, clothed from top to toe in living green, the graceful boles of the white birch, the most abundant tree, gleaming like pillars of silver against the dark background. On the Siberian shore the banks are less bold, but seem to present an endless stretch of rich bottom lands, where the cattle of the world might graze.

One important and interesting feature on all Amour River boats is the constant heaving of the lead, to tell of the treacherous shoals which, all up and down the length of the river, lie in wait for every passing steamer. Constantly, day and night, week in and week out, stands a man near the bow with a stick about ten feet long, loaded at the lower end. To the upper end of this pole a cord is attached, which is tied to the leadsman's wrist, and this pole he constantly casts, once a minute on the average, to ascertain the depth of the water. The pole is divided into foot lengths, each painted a different color, so that he can tell at a glance how much water there is when the leaded end strikes bottom. When the water is deeper than his pole, he gives no warning cry, but when it comes down to nine, eight, seven, six feet, he calls out the depth with every cast, to the man at the wheel, and throughout half the day and half the night we hear his monotoponous sing-song: "shest" (six), "shest polovina" (six and one-half), "pyat" (five), "pyat polovina" (five and one-half). When the leadsman calls out "pyat," O helmsman, look out, for there is only five feet of water below your water-line, and when he calls out "chiteery polovina," ring "half-speed" down to the engine-room, for there is only four feet and a half of water, and the *Baron Korff*, flat bottomed as she is, draws four feet.

These frequent shoals add a spice of excitement to the voyage, for one never knows when he may hear the bottom grating on the gravel, and find the steamer

A THUNDER-STORM BY NIGHT ON THE AMOUR

laid up for an hour or a day or a week on a sand-bar. The Amour River, like certain men, spreads itself out too thin, and gains breadth at the expense of depth.

The cry of the leadsman reminds one of Mark Twain's days on the Mississippi, only they "mark dva," and "tree," and "chiteery," and "pyat," instead of "mark twain," and "mark three," and four, and five. When will the genius of the Amour come to the front? And when he comes will he call himself "Dva" or "Pyat"?

The second and third days on the *Baron Korff* were much like the first—bright, beautiful, rare days in early June. The breath of the flowers came sweet and fresh from either shore, as the steamer swept from one side of the river to the other. No grand scenery marked these early days of the voyage, but much that was exceedingly beautiful, with a quiet loveliness all its own. The whole country has a peculiarly finished and cultivated appearance for such a wild and almost uninhabited land. Though there are few settlements and fewer farms visible from the steamer, there is little of the tangled wildwood of our American forests. Much of the country is open and treeless, and, where trees abound, they grow in stately ranks, with little undergrowth and no impenetrable thickets.

At night the helmsman steers by lights set along the shore at advantageous points and at a distance of a few hundred yards from each other. These lights are like kerosene street lamps, set on posts fifteen or twenty feet high. Some of the lamps are white and others red, and, though their light is feeble, they afford points to steer by, and boats are able to steam day and night, except when stuck upon a sand-bar.

A familiar sight towards evening is the lonely lamplighter in his little Rob Roy canoe, the paddles flashing alternately on either side in the evening sun. Often he is the only inhabitant along the shore for miles and miles, and he must make his solitary rounds alone day after day to fill and light his lamps, and is often obliged to scale almost inaccessible precipices to get at them, for, wherever it is possible, the lamps are placed upon a bluff or headland.

The voyager on the Amour has reason, indeed, to bless the lonely lamplighter, and to say to himself, as the steamer ploughs along through the darkness, guided by the minute points of light in the distance: "How far the little candle throws its beams!"

VII

BETWEEN TWO GREAT NATIONS

FOR three days longer we continued to steam on the *Baron Korff* between China and Russia. On the third day out from Khabaroffsk the scenery grows more charming with every mile. The Amour here passes through a cleft in a spur of the Khingan Mountains, which close in on her on both the Manchurian and Siberian sides. The hills here are not very high, perhaps not more than five hundred feet, but their sides are often precipitous and present a sheer wall of rock to the view, through which it forces its way with many a swirl and eddy and with tremendous force of current, against which our good ship could make but slow headway.

We here remarked again the peculiar

ON THE BANKS OF THE AMOUR

open, park-like effect of the forests, especially on the Manchurian side. I could scarcely believe that these trees in these far Eastern wilds had not been planted by man, and that they did not indicate some great baronial estate, whose castle, to be sure, was hidden behind the hills. Some beautiful hill slopes were covered with stately white birches, without a sign of underbrush, but with green grass forming a soft carpet beneath. In other places trees with black boles took the place of the white birch, but they were at such a distance that I could not determine the species, though they looked like oaks. For miles and miles these natural parks continued, and even where the park-like effect was not so pronounced and beautiful, the lovely open character of the forest continued, for nearly the whole of our fifteen hundred miles on the Amour. The cause of this beautiful effect I do not know. Perhaps in the severe Siberian winters only the fittest survive. The small wood, vines, and parasites of all kinds are killed off by the cold, and only the hardier and stronger trees remain. Whatever the cause, the resultant parks make the banks of the Amour singularly attractive during the brief summer.

The approach to a village was a matter of much interest, both to those on ship and those on shore. When within a quarter of a mile the captain would blow a tremendous blast on the whistle to summon every man, woman, and child to the banks. They would all respond with promptness and despatch, and come streaming down the bluff to the shore, each woman hugging two or three bottles of milk or carrying a pail of butter or a basket of eggs or a bowl of sour cream or a great loaf of black bread, with a hole in the middle like a huge doughnut. Sometimes a woman would appear holding a large goose in her arms as tenderly as though it were a baby, the body of the goose being concealed in the blouse that covered her ample bosom, while its long neck craned out, exhibiting as much interest in the passing show as any one on the shore.

The passengers on the ship would all congregate on the upper deck, many of them with empty bottles in their hands, to exchange, with ten kopeks added, for the full bottles on the shore. There are, of course, no docks or piers at these villages along the Amour, and to make fast to the shore in this swift and shallow water is a work of time and patience. First the anchor is thrown out about a hundred yards from the shore, and cable enough is paid out to allow the boat to drift within twenty feet of the bank. Then a bowline is carried ashore in the ship's boat and made fast to a tree or a post. A stern line is treated in the same way, and by this time the steamer is near enough for the long gangplanks to be pulled ashore. A wooden horse is put underneath to steady them, and then the eager passengers stream ashore, and there is, for a time, a lively barter in milk and eggs and curd cheese and black bread, until the market is exhausted or the passengers' wants satisfied.

The villages on the Amour all have a striking family resemblance, and usually consist of one or two streets of log-houses, often chinked with moss, one or two stores of general merchandise, and, if the village is of any size, a handsome Greek church, and a triumphal arch to indicate that the czarewitch, in his famous journey to the far East, stopped there. Sometimes the arch is anything but majestic in its proportions, a mere pitiful little affair of unpainted wood, ten feet high, but it indicates the loyalty of the Siberians and their love for their "Little Father" just as well as though it was a magnificent and imposing work of art. The larger towns, like Vladivostock, Khabaroffsk, and Blagavestchensk, all have lofty and imposing arches, blazing with colors or shining with burnished copper, as becomes the larger and wealthier municipalities.

Just at nightfall on the third day out

SOME OF OUR FELLOW-TRAVELLERS. SIBERIAN EMIGRANTS ON THE AMOUR

we passed a Chinese village on the Manchurian shore, called into being by some rich gold-mines in the vicinity. It is only placer-mining that has yet been attempted, and the quartz awaits the enterprise and the machinery of the Western nations.

On the fourth day we passed the Chinese town of Aigun, a city of forty thousand inhabitants, and noted, as I have before said, as the place where the treaty was signed in 1868 which gave all eastern Siberia to Russia.

These places, even from the steamer's deck, mellowed and sweetened as they were by distance, looked dirty and wretched and squalid, and showed none of the thrift, neatness, and prosperity of the towns on the Russian side. These were the only Chinese towns that border the Amour for a thousand miles from Khabaroffsk. Besides them we scarcely saw a hut or a wood-pile in Manchuria. Here are vast stretches of virgin soil, rich and productive, millions of acres of noble forest, great park-like estates, laid out by nature, without a single inhabitant. Yet there are uncounted millions of Celestials huddled together in hundreds of cities and thousands of reeking villages in China, who need this sweet air and the wide freedom of these Manchurian fields, which for centuries have belonged to the "Middle Kingdom." That the Chinaman has not occupied his own domain is the more remarkable, for he is in a way an enterprising individual and often leaves home to seek "fresh fields and pastures new." He finds his home in America and Australia, in the Sandwich Islands and Manila, in the Straits Settlement and Colombo, in Japan and Corea—even across the Amour into Russian territory he has wandered in large numbers, while much of his own country is still a wilderness. I suppose the fact is that the Chinaman knows a good thing in the way of a stable and just government when he sees it, as well as the rest of mankind, and that he does not propose to de-

SOME OTHER FELLOW-PASSENGERS. RUSSIAN OFFICERS AND THEIR FAMILIES

velop a country where he will be robbed and pillaged under the form of law, and the fruits of his toil taken away from him as soon as they are ripe.

Almost every day our steamer had to take on a fresh supply of wood, for she had a vast appetite for birch and pine. Fortunately there was enough to satisfy the most voracious of furnaces. When wooding-up time came at night, it was a most picturesque sight. Two great flares would be built of large logs on either side of the gangplank, a cord of the best wood being piled on at one time to light up the scene. Then the roustabouts from the steamer would tumble ashore, each with a long, stout pole and a strap to go over his shoulders. Two roustabouts, working in partnership, would then lay down their poles and pile the heavy sticks upon them till they had as much as they could carry. Then, hitching the straps over their shoulders, they would hoist their load hip high and rush it on board. The banks down which they had to carry the wood were often steep and slippery, and the gangplank was long and unsteady, and there was many a slip and a spill between the wood-pile and the steamer's hold. But these minor accidents made the scene all the more interesting from the passengers' point of view. How the roustabouts looked at the matter I cannot tell. At any rate, they good-naturedly made the best of it, without any profanity that English ears could understand, picked up themselves and their wood, and started on again.

The fourth and fifth days on the Amour were cold and rainy, but we comforted ourselves for the discomfort of the weather by remembering that the rain would raise the river, which was exceptionally low, and perhaps enable the *Baron Korff* to reach Blagavestchensk.

On the afternoon of the fourth day we reached a little village, which consisted mostly of wood-pile, and here the wait continued hour after hour. No one knew the cause of the delay, and the Russian

A TRIUMPHAL ARCH IN HONOR OF THE CZAREWITCH'S JOURNEY THROUGH SIBERIA

of none of the foreign passengers was equal to the strain of finding out. At last it was rumored that the *Baron Korff* could go no farther, and that we must wait for a boat of lighter draught to take us up to Blagavestchensk. After several hours, two other steamers came sweeping down the river, and brought up with a graceful curve and flourish, like a stage-coach in the olden times, at our anchorage. They were little boats, cramped and dirty as compared with the *Baron Korff,* and many were the groans of the passengers that the change must be made. But low water is no respecter of persons, and we were told that at four o'clock in the morning we must make the transfer. At midnight, however, the captain changed his mind and concluded to scrape through if he could, so, getting up steam, at two o'clock he started. The critical shallows were reached. The *Baron* scraped and grated on the sandy bottom. A half inch less of water and she would have been stranded, perhaps for a day. But the captain whistled down the speaking-tube for full speed ahead, the engineer opened the throttle-valve, the engine responded, and, with a mighty shove and jerk, our great leviathan pushed herself over the shallows and into the deeper water beyond. It was a "close call," but on the Amour, as elsewhere, an inch is as good as an ell.

About eight o'clock in the evening, precisely five days after leaving Khabaroffsk, the beautiful spires of the great churches of Blagavestchensk appeared in sight above the tree-tops, and, sweeping around a bend in the river and heading due north, we were soon alongside the bustling wharf of the metropolis of eastern Siberia. In these five days from Khabaroffsk we had journeyed nine hundred and eighteen versts, or about six hundred miles, having made an average of less than five miles an hour, which is about the average rate of speed on the Amour River when there are no extraordinary delays. Here ended the first stage

A MODEST TRIUMPHAL ARCH AT A LITTLE SIBERIAN VILLAGE

of our river journey, for now we must wait two days and then change to a smaller steamer that can more easily scrape over the shallows that are before us on the upper reaches of the river.

VIII

A SIBERIAN CAPITAL

BLAGAVESTCHENSK, a mouthful of a name for any one but a Russian, is a surprising city to be found in the heart of the wild woods of eastern Siberia.

Almost immediately after our departure it was besieged by the Chinese and became well known the world around as one of the storm-centres of the Yellow Tempest. But I must confess that my knowledge of geography was so defective that I had scarcely heard the name of this city three weeks before I saw it.

Had I been particularly conceited on this point, my geographical pride would have received a severe blow, for here is the natural metropolis of a vast section, as large as all New England, with New

"EACH WOMAN HUGGING TWO OR THREE BOTTLES OF MILK"

York, Pennsylvania, and New Jersey thrown in. Here is a city of nearly 40,000 inhabitants, with wide streets, as fine blocks for its leading banks and stores as Portland, Maine, or Portland, Oregon, could show, or any of the smaller cities of the Union; very comfortable hotels, and at least five really fine churches.

The glories of Blagavestchensk, moreover, are not in the past by any means, but largely in the future, as is evidenced by the amount of new building going on. Piles of brick, masons and hod-carriers, and carpenters and stone-cutters, are seen everywhere, busy at their trades, and it is evident that Blagavestchensk is preparing for the time when the Siberian Railway will be completed from Stretinsk to Khabaroffsk, and still further add to her importance as the chief city on the Amour — indeed, the chief city between Lake Baikal and the Pacific.

We did not have time, however, to make these observations on the evening of our arrival, for before the *Baron Korff* was made fast it was half-past nine o'clock, and nearly dark. Having heard of the Central Hotel, and knowing not whether it was good, bad, or indifferent, we directed the drosky man to drive thither, which he did, and soon set us down in front of a large brick building of imposing dimensions. But here our troubles began. No office could be discovered. No one seemed to be in charge. The register which hung in the front hall—a large blackboard with the numbers of the rooms written in chalk—indicated that every room, with one exception, was full. At last a peasant, in Russian blouse and top-boots, appeared who seemed to belong to the establishment; but, alas! he spoke a word of naught but Russian. To polyglot inquiries: "Do you speak English?" "Sprechen Sie Deutsch?" "Parlez-vous français?" he responded with a "nyet," and a helpless shake of the head. At length his instinct, or our bad Russian, enabled him to understand that we wished lodgings for the night, and he showed us to a large room with a single narrow bed, a sofa, and some chairs.

But there were four in this particular party that sought the shelter of the Central Hotel—two ladies, a small boy, and a man. It was evident that all could not be accommodated in these limited quarters. What could be done about it? All the other hotels were crowded. It was ten o'clock at night, and we were on the sidewalk outside of this thoroughly Russian hotel, with our baggage unloaded from the tarantass. At this opportune moment our good genius came to our rescue in the shape of a young Russian lieutenant who spoke French, and who insisted on my sharing his room with him, so that the ladies and the small boy might have the only vacant room. Moreover, he acted as translator, asked for another bed, arranged prices, and in every possible way became our guide, interpreter, and friend. I shared his room for two days, and when I tried to pay for my part of it he absolutely refused, saying, with the utmost delicacy, but with firm insistence: "I could not think of it. You are my guest."

Would such graceful and delicate hospitality be found in New York, or Chicago, or London? I very much doubt it, or whether the average American, or even the exceptional American, would have taken a foreigner into his room, and, for a couple of days, concerned himself for the welfare of a party of absolute strangers.

Our good angel in military clothes had recently graduated at the Russian West Point, and had been stationed at Nikolskoie, on the Siberian frontier. He was on his way to visit his mother at St. Petersburg, and thence to the Paris Exposition. As I expected to travel farther with him, I neglected to learn his name. But he was detained at the last moment, and I shall not be able to send him my acknowledgment of his kindness by post, and I can scarcely hope that he will read

THE LEADING HOTEL, BLAGAVESTCHENSK

them in this account of our travels. What I think will please him most, however, could he but see it in print, I will record here, that such kindness and delicate consideration I believe to be characteristic of the great people of whom he is such a worthy representative.

While we are still at the hotel, it may give some idea of Siberian hostelries of the better class if I say that the rooms at the Central were large and comfortably furnished, but there were some peculiarities which strike a traveller as odd. A bed with a mattress upon it is provided, but no blankets, sheets, or pillows. A common wash-room, not over tidy, served all the guests, and such a thing as a bath-tub or a bath-room seemed unknown. Every guest is expected to bring his own bedding, towels, and soap. Not knowing this, we naturally asked for sheets and blankets and pillows and towels, which, after some delay, were brought, though only one sheet, blanket, and pillow apparently could be raised.

What was my surprise to have a lengthy bill brought when I came to leave, for I was conscious of having had but few "extras." When I inquired for a translation of the items, the waiter pointed to the different articles ordered the night before, and then to the cabalistic words upon the bill, and I learned that the charge was: Blanket, 25 kopeks; sheet, 20 kopeks; pillow, 30 kopeks; towel, 15 kopeks. No wonder the bill was a long, if not a large, one. The room itself was only two roubles and a half, or one dollar and a quarter, for a large double room, and the *table d'hôte* meals, which consisted of soup, meat, dessert, and tea, cost 80 kopeks (40 cents). Breakfast was served in our rooms, and normally consisted of tea made on the spot in a porcelain teapot, with water boiling hot from the samovar, very nice and refreshing, with lemon as a substitute for milk, and bread-and-butter. We did not attempt the coffee, which is not the national drink, and does not compare with the tea. On tea

THE CEMETERY AT BLAGAVESTCHENSK

and bread-and-butter the Russian would be sure to break his fast, though eggs might be had by the Outlander, if he should insist upon it.

It so happened that the day after our arrival at Blagavestchensk was the emperor's birthday, and all the city was in holiday mood. But it seemed to take its pleasure seriously, if not sadly, as the English are said to do. A few flags were flying, the churches were open, and the stores were closed, and a general air of solemnity pervaded the town. The great German store of Kunst & Albers was religiously closed in front and the curtains drawn, but I was told that it had a back door, like the saloons in Maine. As my conscience did not compel me to observe the emperor's birthday sacredly, I sought the side entrance, where I made some necessary purchases.

I may mention for the benefit of future travellers that many of the necessaries of life cost about the same in eastern Siberia as in New York—linen collars two dollars a dozen, cuffs twice that sum, shirts, underclothing, and other dry goods in like proportion. Two of the stores in Blagavestchensk, Kunst & Albers and a Russian store nearly opposite, are really palatial in proportions and equipment. The department store has blossomed out in full proportions in Siberia, and it is far more surprising to find such a store as I have mentioned in Blagavestchensk than to find Wanamaker's, in Philadelphia, or Jordan & Marsh, in Boston.

Here you find books and bric-à-brac, pipes and pickles, flowers and flat-irons, oil and oysters, waistcoats and watches, indigo and icons. Here were attentive clerks and cash-boys (Chinese cash-boys, by-the-way), cashier girls in their little elevated desks, receiving the money, and all the conveniences and appliances of the great modern establishments.

But the chief interest in Blagavestchensk centred in the churches on the day of our visit, especially in the blue and green military church. Here a high

ONE OF BLAGAVESTCHENSK'S BEAUTIFUL CHURCHES

mass was held in honor of the emperor. The church was crowded with soldiers and peasants, and all were most devout and earnest. Old peasants would come in, and, though there was scarcely standing room, would spread down their bandannas, and, kneeling on them, would touch their grizzled forelocks to the floor, and then rise and join heartily in the service. This was conducted chiefly by a priest in beautiful robes of cloth of gold, whose magnificent bass voice chanted the service as I never heard it rendered before. A sweet-voiced choir of boys answered the priest, and the whole congregation joined reverentially, but in stentorian tones, in part of the service. Behind the altar and facing the officiating priest, whose back was to the audience, stood a figure clad, apparently, in pure gold, and with a golden mitre upon its head. Whether man or image I could not at first make out. But it stood so immovable, and was apparently so solidly built of gold, that I concluded it was an image, a huge icon, which did duty in the soldiers' church. I thought that I detected the image solemnly winking, but the moment after I concluded that I was the victim of an optical illusion. Nothing so stately, solemn, and immovable could wink. It was almost sacrilegious to think of such a thing. But when I had fully made up my mind to the icon theory, the golden image stepped from behind the altar, and, coming to the front, an attendant took off his heavy mitre, and in a deep, rich voice, he intoned the part of the service that fell to his share.

Bowing to the cross, his golden garments wrinkled and bent like the scales of a fish, or like the chain-armor of an ancient knight whose corselet was of pure gold. Never have I seen in any church or temple, in Christendom or pagandom, such gorgeous vestments or a more stately service. I only hope that it was as intelligently devout as it was gorgeous and impressive.

Some of the Greek priests, whose writ-

A GREAT DEPARTMENT STORE IN BLAGAVESTCHENSK, WITH MARKET-PLACE IN FRONT

ings I have read, like Father John, of Kronstadt, are as deeply imbued with the Spirit of God as Thomas à Kempis or John Tauler of old, or as Andrew Murray and F. B. Meyer of the present day. A fine, sweet, mystical vein runs through their writings, which shows that they have seen the Invisible and heard the Inaudible, and their writings are as refreshing as a draught of clear spring water in this thirsty commercial age.

So may it have been with the holy fathers of Blagavestchensk! I could not judge, for all their service was to me in an unknown tongue. I will give them the full benefit of the doubt.

The streets of Blagavestchensk on this holiday were full of soldiers, who marched in orderly file from their barracks to the church and back again. They were fine, stalwart, manly fellows, marching in perfect alignment, and with an easy swing which did one's eyes good to see, after beholding for a few weeks the slouchy straggle of Chinese soldiers. They were an orderly, well-behaved lot, too, and I saw no rioting or drunkenness. The log-cabin which served as a soldiers' guard-house seemed to have few occupants. The soldiers did not know how soon their holiday manners would be changed into grim and deadly warfare with the yellow men on the opposite bank of the Amour.

Blagavestchensk, though it has a few large and imposing buildings, is largely a log-built city. Nine-tenths of the houses are of hewn logs, but most of them are prettily painted and are not without appropriate ornamentation. I never had believed that a log-house city could be so picturesque and substantial in appearance.

Upon the churches the people have lavished their chief wealth. At least three of them are costly and beautiful buildings that would do honor to any European capital, and a fourth that is building will, when completed, surpass them all. It is really a building of remarkable stateliness and beauty.

AN UPPER AMOUR RIVER STEAMER

Buriats' hut in foreground

In the cathedral I saw for the first time the Black Christ and the Black Virgin, which are often seen in Greek churches. The pictures in all the churches impressed me as a much higher grade of art than one sees in the ordinary church on the continent of Europe.

When we were in Blagavestchensk we had little thought that this obscure capital was so soon to figure in the world's telegrams, or that it could be besieged and bombarded by the Chinese. But history in the far East was making itself rapidly about that time, and we had scarcely left Blagavestchensk a week behind, when the apparently peaceful Celestials, inspired from Peking, attacked and for a long time bombarded this inoffensive city.

As I write, the outcome of the immediate conflict is still in the balance, but it can scarcely be doubted that in the long run the Giant of the North will win his way through Siberia and Manchuria into the heart of China, and that the siege of Blagavestchensk will really hasten the awakening of China from her sleep of the ages.

On the evening of the second day, we were glad to take passage once more for the upper reaches of the Amour, the most difficult, uncertain, and tedious part of our journey.

"*If you have tears, prepare to shed them now*" might have been our motto as we went aboard the crowded little steamer and realized that Stretinsk, the terminus of our river journey, was 1197¼ versts away, and that, much of the way, less than four feet of water separated our keel from the sandy bottom of the Amour. But how we scraped over the sand-banks from Blagavestchensk to Stretinsk must be reserved for "another story."

ON THE UPPER AMOUR

IX

AGROUND IN THE AMOUR

TWO Mississippi River captains were once conversing as to the merits of their respective steamers as light-draught boats. One claimed that his steamer could run in a good heavy dew. The other, not to be outdone, claimed that that was nothing, for his boat could run on the sweat of an ice-pitcher. It is this latter kind of boat that is needed on the upper reaches of the Amour River, as we began to think before we were many miles from Blagavestchensk. We had scarcely gone five miles before we heard that peculiar crunching, scraping, gritty sound which we came to know so well. The paddle-wheels revolved fruitlessly for a few minutes, but it was no use, and we knew that we were hard and fast aground in less than three feet of water. For hours the boatmen worked with might and main, and at last we could feel that there was water under our keel and that we were making slow progress against the current once more. But we had scarcely been moving another hour when c-r-u-n-c-h, s-c-r-a-p-e, s-t-o-p, and there we were stuck on another sand-bank. The captain seemed to tear his hair in despair, and well he might, for Stretinsk was 1187 versts away, and we had scarcely come ten versts in ten hours from Blagavestchensk. If it requires ten hours to do ten versts, how long will it take to reach Stretinsk, was an appalling problem in mental arithmetic.

However, we made somewhat better progress after getting off the second time, and the hope that springs eternal in the human breast kept us from despair of reaching in time the important engagement at our journey's end.

The *Ivan Vishnegradskie* was a much smaller steamer than the *Baron Korff*,

TUMBLING ASHORE FROM THE *ADMIRAL CHICACHOFF*

and consequently much more crowded. Especially were the third-class passengers much in evidence, for there were no steerage accommodations, and they were obliged to lie about anywhere as best they could, in the passages, on the companion-way, huddled on the boiler deck, lying out on the wood-pile or anchor chains; every available foot of deck room had its two human feet to occupy it.

It is said that the birth-rate of Siberia is higher than any other country in the world, amounting to fifty-six in the thousand. I could well believe it as I saw the uncounted, tow-headed babies tucked into every nook and cranny of the *Ivan Vishnegradskie*. Head to heel and heel to head, they lay, with their fathers and mothers and older brothers and sisters lying about them four square, building, as it were, a sort of living rampart around the family compound. It was pathetic to see a mother lay her three little children out in a row, with the croupy baby in the middle, while she herself sat in the draughty doorway all night long to shield them from the wind.

"She bared her bosom to the storm,
And smiled to think her babe was warm."

For two days we sailed on this crowded little craft. The river turns almost due north after leaving Blagavestchensk. A very tortuous stream is the Amour from here on, turning and twisting and doubling on itself like a huge serpent writhing in agony. Following its twistings we made very little progress, even when we could steam all day long, which the treacherous shoals did not often allow us to do. The shores on either side were often picturesque and beautiful, and sometimes grand and striking, with bare cliffs of rock rising sheer from the water's edge to a height of four or five hundred feet.

The park-like effect of the clean, open woods continued, and made it difficult to believe that the country had not been thickly inhabited for a thousand years, except that the utter absence of habitations dispelled the illusion.

On the second day out from Blagavestchensk we were obliged to change steamers and take up our abode in the *Admiral Chicachoff*, the regular steamer of this line, which had not been able to get so far down the river as the above-named city. The *Admiral* was a somewhat larger boat than the *Ivan*, and had much better accommodations for the steerage passengers, who no longer had to sleep in doorways and draughty passages, but had the after upper deck roofed over for their accommodation. The steamers of the line on which we were embarked carried the government mail, and were supposed to be more sure of reaching their destination than the other river-steamers, of which there are many lines on the Amour.

Frequently we passed other steamers going down with the current at a famous rate of speed compared with our slow and toilsome ascent of the river. More often we passed great rafts of timber, with little bark huts built upon them, in which families lived and moved and had their being for weeks and perhaps months at a time, as they let the force of the current carry them down the stream, and tied up to the bank at night to avoid the passing steamers. Sometimes the rafts would be loaded with horses that seemed to enjoy their floating quarters, and who lived off the country, for it was only necessary to tie the raft to the shore and drive its living freight into the rich pasturage, which in many places could be obtained free, gratis, for nothing.

The monotony of the voyage was varied by a stop once or twice every twenty-four hours to take on wood, or at some little log village on the river's margin. On the occasion of these stops the passengers would tumble ashore almost before the gangplank made connections, and would spend the time as the fancy seized them, in the woods or on the pebbly margin of the river. The youngsters would scale the bluffs, which usually abounded near

CHILDREN OF EMIGRANTS PICKING WILD FLOWERS ON THE BANKS OF THE AMOUR

the wood-piles, or would come back from the meadows laden with great armfuls of beautiful wild flowers. Lilies of the valley seemed to abound everywhere, violets, blue and white forget-me-nots, lady slippers, and another beautiful orchid somewhat similar, which I have never seen elsewhere. In other places, blue honeysuckles were found, and yellow lilies, and Solomon's seal, and the yellowest of buttercups. Indeed, the flora of the country seemed to be exceedingly extensive, embracing most of the spring flowers with which I am familiar in New England, except the arbutus, and many flowers that I had never seen before. The trees were mostly white and yellow birch, poplar, white oak, soft maple, larch, and pine. The wood-piles, cut for the use of the river steamers, were interminable, and consisted of beautiful, straight, knotless sticks of birch and pine that it seemed a pity to sacrifice in the rapacious maw of our never-satisfied boilers, to go up in smoke and sparks.

While the smaller fry were racing on the shore or gathering wild flowers, their more sedate elders among the steerage passengers would kindle a little fire of driftwood on the shore and cook a savory pot of stew to vary the crude monotony of their daily fare. Or the forethoughtful mothers would improve the opportunity to scrub their offspring or wash their clothes, pounding the latter upon the stones. Sometimes we would get a delicious bath in the clear, cold water which, less than a month ago, was bound with icy fetters, for the weather, much of the time, was extremely warm. The summers of Siberia are short and fiercely hot. Vegetation has to be forced to maturity by Dame Nature in six weeks, instead of taking six months, as she would do in more leisurely climes.

Many patches of snow and ice still lingered upon the banks, and it could have been comparatively but few days since all the country was snow-bound. Yet every tree seemed to be in full leaf, and ferns

AN EMIGRANT HOUSE ON THE UPPER AMOUR

and grasses in many places were knee-high and were growing with the lustiness and vigor of a tropical summer.

Our new steamer was no more exempt from the prosaic perils of the sand-bar than the one we had just left, for, before we had finished the first day we had run hard aground in a place where the current was very swift, and where it was very difficult to get off. For seventeen hours our officers and crew worked with a will, but to no effect. First huge poles, like iron-shod battering-rams, as big as telegraph poles, which could barely be lifted by six men, were hitched to the windlass by a wire rope, and, with this, every effort was made to pry us off. But all utterly failed. Then two telegraph poles were rigged, and finally three, but all in vain. In the useless effort, three or four of these huge pries were broken off short, and yet the little steamer did not budge. Then, to lighten ship, the convict barge, which all these days had been following in our wake, was towed alongside, and all the passengers were requested to go on board of her.

This gave us an opportunity to visit our fellow-passengers behind the bars, if I may so call those who were propelled by the same steam and were hitched to us by a single cable. It must be confessed that they were not bad-looking fellows as men average, and the impartial visitor, had he been asked to pick out the convicts by their looks, would, very likely, have chosen the hardened criminals from among the free men on top of the barge, instead of from among those whom the law had put into the cage below. Possibly even some of the first-class passengers who had paid thirty roubles for their voyage would have been chosen, instead of those who had fare and board paid by the government. My idea of the severity with which Russia treats her Siberian exiles was also modified when I saw the clean and comparatively comfortable quarters of the prisoners, which compared most favorably with the steer-

THE MEN BEHIND THE BARS ON THE PRISON BARGE

age accommodations on our own steamer. There were not many prisoners in the cage, and they were going from the crowded prisons of Vladivostock or Khabaroffsk to other prisons in the interior. One poor fellow was landed at a little river town, and in his heavy felt boots and overcoat, though it was a broiling day, was marched between two soldiers to a new prison, or, perhaps, handed over to some man who had bought his labor for a time.

But we had gone aboard the convict barge to lighten our ship, and not to inspect the prisoners. It was all of no avail, however, and some new method must be tried. So an anchor was carried out into mid-stream, two hundred yards away, perhaps; the steam windlass was again brought into play and, little by little, inch by inch, we were coaxed off the sand-bar, the flat bottom creaking and groaning as the steamer scraped along, as though unwilling to leave her comfortable bed. Seventeen hours had been wasted, and not a third of our voyage had been accomplished.

After this experience our captain was very cautious, and when he came to shallow places would anchor, and often spent several hours in a small boat in ascertaining the channel and marking it with sticks driven into the sand. Then the *Admiral* would start up slowly and warily. Two men, one on either side of the bow, with sounding-poles, would constantly take the water and report to the first officer of the deck above, who shouted it back to the man at the wheel. "Vosem" (eight), "sem" (seven), "shest" (six), he would call, and no one would feel anxious, for the *Admiral* only draws two feet and a half; but when it came to "pyat polovina" (five and one-half), to "pyat" (five), or to "chiteery" (four), we began to be interested.

When, in stentorian tones, the leadsman would call "trees polovina" (three and one-half), and his mate on the other side would call "trees," and the first officer would shout to the wheel, "trees," we would positively hold our breath, and sit more lightly in our chairs to ease the ship, until again the leadsman would reassuringly cry "chiteery," "pyat," "shest," "sem," and we would feel that we were safely over one more bar.

X

ON THE SHILKA

OUR progress up the Amour was slow but not sure. Time after time the experience described in the last chapter would be repeated. We would touch bottom, scrape on the sand, the engines would make desperate efforts to pull us off into deep water, but would finally give up in despair, slow down, stop altogether, and for one hour, or ten, or twenty, we must possess our souls in patience, until, with infinite pains, our craft was floated once more.

We were told before starting that the journey from Blagavestchensk to Stretinsk would not take more than seven days, or seven days and one-half; but seven times twenty-four hours passed and Stretinsk was still four hundred

THE HILL-SIDES WERE CARPETED WITH FLOWERS

miles away, and not half the distance had been traversed. The alarming tale was circulated that the water was falling at the rate of two inches a day. The usual story about this being an extraordinary season was current—"not for thirteen years had the water been so low," etc. I am inclined to think that every year on the Amour is "extraordinary," and that, especially in the month of June, those who essay this journey must count on spending as much time on the sand-banks as in deep water.

Still, for those who are in no haste to reach their journey's end, who can put up with very indifferent food, and who can good-naturedly accept the inevitable, there are many and delightful compensations in this new way around an old world. The scenery throughout the whole journey is pleasing, and, much of the way, it is extremely beautiful. A hundred miles beyond Blagavestchensk we passed some remarkable high bluffs, seamed with coal or lignite in narrow, diagonal veins. In many places this coal had caught fire, and for years had been burning with an unquenchable flame. The scene, as we passed, was one of striking beauty. The sun was just setting behind the western hills, lighting up the evening clouds with golden and crimson glory. The broad river swept in a grand and graceful curve around a pine-clad mountain that towered a thousand feet above its placid waters, while, as the light of the dying day waned, glowing jewels appeared in the mountain-side, where the burning coal was afire, or little bursts of flame, as from a hundred gas-jets, lit up the gaunt and seamy cliff.

Two days later, and a hundred miles farther on, the same scene was repeated with a still lovelier variation. Again it was just at the hour of setting sun, the most delicious of all the day, for the boiling heat of the day was over, and the cool breezes of evening had begun to blow. On the Chinese shore of the great river high cliffs like the Palisades

A TYPICAL WATER-TOWER ON THE TRANS-SIBERIAN RAILWAY

of the Hudson stretched westward for miles, with tops as even and flat as Table Mountain at Cape Town. Here and there, throughout their whole length, these cliffs, too, were seamed and veined with fire. Oftentimes green shrubs overhung the seams of coal, or pine-trees grew below them, their upper branches athwart the narrow line of fire, so that again the Scripture miracle seemed to be repeated, and the burning bush was not consumed. Imagine the Palisades on a wilder and grander scale, illuminated by their own internal fires, and you have an idea of this night scene on the Amour.

Seven days out from Blagavestchensk, the day we should have completed our river journey, we again stuck fast on a sand-bar for nearly twenty hours. This delay was all the more uncomfortable on account of the extreme heat, unalleviated by the breeze caused by the boat's motion, and the discomfort was aggravated by innumerable huge green-headed flies, larger than the biggest horse-fly I ever saw in America. These flies showed all the idiosyncrasy of the average house-fly in returning time after time, with the utmost persistence, to the same unguarded spot on ear, or nostril, or back of neck, and, being about ten times as large, their attentions were fully ten times as unpleasant as those of the house-fly.

At length we rubbed our way, inch by inch, over the shallows and into deeper water, and an hour later found ourselves at Albazin, a small and decayed village four hundred miles from Blagavestchensk. Albazin has no glories in the present, but it is famous as the place where, two hundred years ago, a Russian garrison was defeated by the Chinese, and four hundred prisoners carried off to Peking. As a result of this victory a treaty of peace was concluded, and the north bank of the Amour again became Chinese territory. The Great Bear of the North is sometimes defeated and retreats to her den, but it is only to nurse her wounds

THE LIGNITE CLIFFS ON THE AMOUR

and to come out again, stronger than ever, to win, either by arms or diplomacy, perhaps two hundred years later, a new victory which will wipe out old scores.

Two hundred years have elapsed since the Russian garrison was carried off to Peking, but Albazin is again in Russian territory, which stretches fifteen hundred miles beyond to the very shores of the Pacific, while Manchuria itself and China, almost to the very gates of Peking, seem likely soon to become Russian provinces.

"That, too, is our country," said a Russian judge, one of our travelling companions, as he pointed with a sweep of the hand to the Manchurian shore. "That, too, is our country." And his was no empty boast, even if a little premature.

It is an interesting fact that the descendants of the Russian garrison carried from Albazin to Peking still live in the Chinese capital. They married Chinese wives, but organized a distinct community of their own, in which the Greek Church has flourished ever since.

Ten miles from Albazin we stopped at the larger and more flourishing town of Renova, that owes its prosperity to some large placer gold-mines, which, like all Russian gold-mines, are a government monopoly. All gold discovered and dug must be sold to the government at a price fixed by it below the market price. Of course, much is smuggled out of the country, and such a law is death to genuine private enterprise; but the laws are strict, and no foreigner is allowed to filch the gold from the czar's private purse, *i.e.*, the bowels of the earth.

The next day we reached the final limit of travel on the *Admiral Chicachoff*. She could go no farther, and we were transferred to a great barge, not unlike the prisoners' cage which we had been towing all this distance. This barge, which for another week was to be our abiding-place, drew only one foot and a half of water, and was towed by a tug which also drew but eighteen inches. On this craft we were to enter the Shilka and stem the current for three hundred miles farther to Stretinsk.

Each change we had made was a descent. We had gone from luxury on the *Baron Korff* through different grades of discomfort, until we had come to the superlative degree of discomfort on the barge *Diana*.

The men among the first-class passengers were all herded together in one cabin, and the women in another, and the decks and passageways were so littered with third-class passengers and their belongings, that it was with great difficulty that we could pick our way over stout, extended legs, and tender baby toes, and peasant women suckling still smaller babies, who filled every nook and crevice. Clad in great sheepskin garments,

> "With the skinny side out and the woolly
> side in,"

and with huge top-boots, worn often by the women as well as the men, the Moujiks defied the cold night air and the intensely hot mid-day sun, always sprawled out, day and night, at full length, except when they went ashore at landing-places to cook their scanty meals over a brush-wood fire.

Of course, exercise under these circumstances was out of the question on shipboard, and, as can be imagined, we all made the most of our stops at the wood-piles for a little run on shore.

Travel on Siberian rivers, like poverty and politics, makes strange bedfellows, and in the men's cabin were a polyglot company, consisting of a German baron, a Swiss consul, a Dane, two French "bounders," two Russians, two Scotchmen, a North of Ireland man, and two Americans. All made the best of the situation except the Frenchmen, who sputtered and growled at everything, and made themselves generally disagreeable. Like Mrs. Gummidge, they seemed to say of all the discomforts: "I feels it more than others." Especially did the food, which indeed was atrocious, come

THE EVER-PRESENT SAMOVAR

in for their objurgations. At last the climax was reached when one of these irascible Gauls found that sugar had been put in his cabbage. His gastronomic canons were all outraged, and, throwing the cabbage, dish and all, out of the window into the river, he flung a full bottle of beer at the waiter, narrowly missing a lady's head, and only then felt that his outraged culinary taste had been fully avenged.

But for the most the company on the *Diana* was a happy family, and though a "Never Again Society" was proposed (never again to cross Siberia by this route), we often consoled ourselves with the reflection that things "might be worse."

On the second day after boarding the barge the Amour, so far as we were concerned, came to an end, or, rather, we came to its abrupt beginning, for, at a junction point of entrancing scenery, the Argun and the Shilka, two great rivers, come together to form the Amour, after having pursued their separate and devious ways for hundreds of miles. The Argun flows sharply from south to north, the Shilka pursues a northeasterly course. To be sure, the Shilka is really the Amour, and ought to bear the same name to its source, but for boundary purposes the Russians choose to call it another river, and to bound their possessions by the left branch of the Amour, called the Argun. We turned into the Shilka and were soon steaming against its swiftly flowing current.

Now, at length, after coasting along the Manchurian shore for over a thousand miles, we leave China and plunge into the heart of Siberian Russia. On either side the mountains close in on the Shilka, dipping their feet in its rapid, crystal stream, and narrowing it down to less than a quarter of a mile in width. Wild and picturesque is the scenery for almost the whole distance to Stretinsk. The fertile arable land and vast open parks of the Amour give way to rough and

SOME FIFTH-CLASS PEOPLE

rugged hill-sides, clad with a growth of small trees, but affording scant encouragement for the agriculturist. The villages are fewer and farther between, and the scanty houses are of a distinctly poorer type. In fact, the whole country is given up to placer gold-mining and to cutting wood for the frequent steamers, and almost no cultivation is attempted.

It even became difficult, and at last impossible, to secure milk and eggs on shore, and some of the scanty fare provided by the steamer's steward might have been called "rotten" by the college boy without indulging in slang.

But the natural scenery grew more charming as the evidence of man's handiwork disappeared. The hill-sides were carpeted with flowers; lilies of the valley, azaleas, and a half-dozen kinds of rare and beautiful orchids abounded everywhere, and the air was fragrant with the new-made leaves of birch and pine. Cuckoos struck all hours of the day from the deeper forest, and a sylvan beauty that I have rarely seen surpassed constantly greeted our eyes.

Frequently during these days on the Shilka we passed great rafts, floating down stream with the current, loaded with emigrant families, men, women, and children, horses, cattle, and dogs, and all the little store of household goods that the poor Moujiks possessed. Sometimes a small bateau would contain a family of seven or eight people and all their belongings. Occasionally great barges towed by a tug-boat would pass us, and these were invariably crowded with such a mass of humanity that one wondered how all could stand by day, much less lie down by night. These crowded rafts and barges headed eastward were most suggestive of Russia's policy. Hers is a peaceful conquest of the far East. First she must fill up the waste places of Siberia, and then she can stretch out her hand over all Manchuria and northern China, and even in spite of temporary Mongolian up-

A RAFT ON THE SHILKA

risings she can hold her own, with no fear of a repetition of the disaster of Albazin, and another exclusion for centuries from Manchuria.

With one barge company alone the government has made a contract to transport thirty thousand peasants within three months to the lower waters of the Amour. Thousands of others will find their way by other methods of conveyance. Liberal land laws and government aid while they are getting established will result in a few years in tens of thousands of log-house homes that in time will make this wilderness to blossom like the rose.

As we approached Stretinsk signs of civilization became more numerous. Cultivated fields appeared. Winter wheat already seemed to be ripening, forced to maturity by a few weeks of the almost continuous sunlight of this northern clime. The interesting village of Oost Kara, about seventy miles from Stretinsk, lies on the right bank of the Shilka, and is supported by extensive gold-mines, which are worked by convicts—criminals, not political prisoners, I am told.

At length, on the 21st day of June, thirteen days after leaving Blagavestchensk, and twenty-one days after embarking on the Amour at Khabaroffsk, we were glad enough to bid good-bye to the *Diana,* its crowded decks, and its wretched fare, and still more glad to feel that our long river voyage, which had stretched itself out from the twelve days which we had expected to twenty-one, was at length at an end, and that the rest of the journey would be behind the iron horse on land, instead of his aquatic cousin, the puffing steamboat.

XI

THE TRANS-BAIKAL

WE had supposed, as has already been remarked, that if we had tears to shed, we should need to indulge them on the last part of our river journey. But we soon found that we should have saved them all for the journey on the Trans-Baikal Railway between Stretinsk and Irkutsk, as the sequel will show.

Stretinsk is a raw, shabby, straggling frontier town, whose only excuse for being so uninteresting is that it is very new and never thought of being a place of any importance until it suddenly found itself, on the 28th of December, 1899, the terminus of the Trans-Baikal Railway, and an important forwarding point for freight and passengers between the

Atlantic and Pacific. The large railway shops and engine-houses are situated across the river from the town, which is connected with them by a ferry, ingeniously worked by the swift current of the Shilka, like the picturesque ferry at Bonn, on the Rhine. The hotels of Stretinsk are wretched inns, and there are absolutely no attractions within the town to persuade the traveller to remain. But, in justice even to the squalid raggedness of Stretinsk, it must be admitted that fine hills encircle it and a noble river washes its feet, for the Shilka, in spite of shoals and shallows, is still a mighty stream, and runs even here with a large, impetuous flow.

We were glad enough to find that we must remain in Stretinsk only thirty-six hours, and that the train which was to carry us to Lake Baikal was to start the following night. We were still more rejoiced to find that the good governor of Vladivostock, General Tchitchagoff, had not forgotten us, but had wired for a

SIBERIAN RECRUITS BOUND FOR CHINA

special car for our use. The car was very small, and held but five people, and, as can be imagined, it was not built on the latest Pullman plan, but, as compared with the cars assigned to our three hundred fellow-passengers, it was a *car de luxe,* indeed.

This road is not yet officially opened for traffic, and we were taken on sufferance, as it were. At least, most of those who take it have to suffer severely. Nor can they complain overmuch, for the government does not invite them to go or advertise any attractions. The best cars that are yet run, with the exception of the little *car de luxe,* of which I have just spoken, are marked "fourth class," and are no better than they pretend to be. They have wooden seats of the hardest possible variety, and three wooden shelves, one above another, afford cramped opportunity for a man not more than five feet long to stretch himself out. Each of these cars, with three tiers of shelves, is supposed to afford accommodations for forty-three people. But these are the best, as I have said. Others, which might be termed fifth class, if the nomenclature of railway trains descended so far, are simply box-cars, with no seats, and marked on the outside, "to carry twelve horses or forty-three men." Into these cars there crowded, helter-skelter, pell-mell, higgledy-piggledy, Russians and Siberians, Moujiks and Chinamen, Tartars, Buriats and Englishmen, Frenchmen, Germans, and Americans.

If there were fifth-class cars, there were plenty of sixth and seventh class people—some in rags, and many in tags, but few in velvet gowns. Old Moujiks, with half a dozen half-naked children, filthy with a grime that has accumulated since their birth, and alive with unmentionable parasites, crowded every car, or, rather, human pigpen, as each car soon becomes. Odors indescribably offensive made the air thick and almost murky. The stench, the dirt, the vermin grew worse the longer the car was inhabited,

FOR THIRTY-SIX MORTAL HOURS WE WAITED

and one simply resigned himself to the inevitable and lived through each wearisome hour as best he could.

Our non-Russian fellow-passengers, when they found the true state of affairs, made a desperate resistance to their fate, and after repeated remonstrances, objurgations, and even threats, which nearly came to blows, induced the station-master to put on another little car, which they could have for themselves.

We fondly hoped that the governor's kindness had spared us the heaviest affliction of the journey; but alas! even the best-hearted governor cannot insure all the bridges on the line, and we had not been travelling a half-dozen hours when we learned that a bridge had burned down twenty miles ahead, owing to the carelessness of a fireman who drew the fire on the wooden sleepers, and there we were stranded as fast as though on a sand-bar on the Shilka.

For thirty-six mortal hours we waited at that wretched railway town, which consisted principally of a round-house and a water-tower, then were pulled to the burned bridge, twenty miles away, and, with much difficulty and long delay, were transferred to the train on the other side. Here our good fortune deserted us. Our comparatively fine car could not, of course, cross the bridge, the governor could do nothing more for us, and in the general scramble we found ourselves in one of those dreadful fourth-class cars I have already described, with the Moujiks, the filth, and the vermin destined to be our companions for four weary days and nights.

After passing the burned bridge our train met with no serious delay, though our rate of progress was exceedingly moderate. The whole distance to Irkutsk is about seven hundred miles, and we managed to consume three days and a half of actual travel in covering it. It would seem impossible to the uninitiated in Siberian travel that a railway train that could go at all could go so

A TRANS-BAIKAL TRAIN. PRISONERS CARRYING WATER

slowly. Two hundred miles a day, or about eight miles an hour, is scarcely more than post-horses would make. This is accounted for by reason of the interminable, unreasonable length of the stops at the stations. Our train would draw up at a wood-pile and a log-house. The peasants would scramble out of the train, build their fires, cook their soup, boil their tea, and still the train would wait. There was usually no baggage to be taken on or put off, no passengers to join us, no passing train to wait for. Water would soon be taken, and still we would wait, a half-hour, three-quarters, a full hour. At last, for no particular reason, apparently, the station-master would ring a big dinner-bell. Five minutes later he would ring another. Then, soon after, the guard would blow his whistle, the engineer would respond with the engine whistle, the guard would blow again, the engineer would answer him once more, and, after this exchange of compliments, the train would move leisurely along, only to repeat the process two hours later at the next station.

The stops ranged from half an hour to two hours in length, and the speed attained about fifteen miles an hour while running. Fortunately the stations are few and far between, averaging nearly thirty miles apart.

The only towns of any importance on this line are Chita and Verkneydinsk, the former about two hundred and the latter about five hundred and fifty miles from Stretinsk. Verkneydinsk is a centre of the old caravan tea trade, and retains its life in spite of the new railway.

The Trans-Baikal province, which this railway traverses from end to end, is the smallest of the six provinces into which Siberia is divided, and has about six hundred thousand inhabitants, of whom two hundred thousand belong to the aboriginal tribes. Much of it seems exceedingly fertile, and vast savannas, stretching away from the railway track as far as the eye can reach on either side, are frequently seen during the early part of the journey. Thousands of cattle and horses roam on these great flat prairies, which, often, are as level as a house floor for miles. In other parts of this journey the scenery is beautiful and picturesque, but never grand, the mountain range which is crossed being not a lofty one. For many miles the railway follows the winding of the Shilka River, usually high above its banks. Then it crosses the plains, and afterwards, some four hundred miles from the starting point, plunges among the pine-clad hills which hem it in on both sides with their resinous heights.

In other parts of the hill beautiful birches stand like white sentries in a green livery. The flowers that carpet the fields everywhere are lovely; marigolds and marguerites, roses and blue honeysuckles, buttercups galore, bold tiger-lilies, modest lilies of the valley, and white and yellow poppies make the trackside gay.

After traversing about half the distance, the streams, which had been flowing east into the Pacific, give way to others that flow west to Lake Baikal. The roof of the world, or, at least, of this part of the world, has been reached, and now we will follow the raindrops towards the distant Atlantic.

The Selenga, which we reach on the last day of this journey, is a great and strong flowing river, as large as the St. Lawrence above the Ottawa. It is crossed by a splendid iron bridge, resting on magnificent piers of cut stone, a bridge which would be an honor to any railway in the world.

It is the custom of the few travellers who have crossed this line, or any part of it, to poke fun at the Trans-Baikal Railway. And, indeed, it is not hard to do so. With its crawling trains, its inordinately long stops, its primitive rolling-stock, it does not inspire much respect. It reminds one of the railway in the United States called a "tri-weekly

BOUND FOR FAR SIBERIA

A passing glimpse of an emigrant train

road," which was explained by its president to mean that a train went up one week, and tried to come down the next.

All the ancient and hoary railway jokes, like the one about the boy who started on a half-fare ticket and was so old before he reached the end of his journey that he had to pay full fare for the last part, are cracked and appreciated by the passengers on this line. Yet it must be remembered that the last spike in this road was driven less than six months before I passed over it; that it was not even then accepted by the government, or formally open for traffic; that it is largely built for military exigencies, and that no one is asked to travel over it, but rather discouraged by Russian officials; then the jokes lose their best points.

Still it must be confessed that the road seems to tithe the mint, anise, and cumin, and omit the weightier matters of railway construction. For instance, the water-towers are beautiful, stately structures, and the stations are very creditable even for an old railway, but the rails are light and constantly breaking and giving way and delaying traffic for days at a time. "Two streaks of rust across Siberia," is the exaggeration of a friend, which has an element of truth in it.

Many parts of the embankment have been carefully sodded, the sods being pegged down with great care, but the road is very imperfectly ballasted, and is rough almost beyond belief. The culverts and small bridges are buttressed with cut stone, carefully dressed; the cars, as I have said, are exceedingly poor and filthy.

Still the road is evidently built for the future, and all these defects will, in time, be remedied, and the Trans-Baikal section will take its place as an important link in the greatest railway of the world.

During these long and weary days we were obliged to get what diversion we could from our fellow-passengers. We

FIFTEEN MINUTES AT EVERY STATION

watched the mother who combed her offspring's head with a carving-knife, with which she made vigorous onslaught on the numerous inhabitants that had taken shelter there, while we shuddered as we thought of the coming night in close proximity. We were interested in the other mother who did her daily washing in a pint cup and hung it out to dry on the upper bunk. We were fascinated by the soldiers who performed their morning ablutions by taking a large mouthful of water from a teapot, then, squirting it on their hands, vigorously washed their faces.

We never get away from the prisoners in Siberia, and two cars, immediately in front of ours, were filled with these poor wretches. Before these cars, at every station, marched four soldiers with set bayonets. The heads of the prisoners, shaved on one side only, would have betrayed them even had they escaped for a little.

At the stations women and girls with

milk and eggs were eagerly patronized by us all, for we were desperately thirsty, and water that we dared to drink was scarce. At some stations we saw the aborigines of the country, the Buriats, of whom two hundred thousand remain in the province. They are a Mongol race, and their black hair and slant eyes contrast with the blue eyes and tow heads of the Siberians, bleached as they were still further by the sun.

Some of the Buriats are won to Christianity, but most of them are rank heathen still. They are a nomadic race, and do not take kindly even to the limited civilization of Siberia.

I must record that, in the midst of the filth and discomfort and unutterable odors of this hard journey, we met with many courtesies and kindnesses from the most unpromising of our fellow-travellers. Some of the peasants were ladies and gentlemen at heart, who would incommode themselves to promote our comfort, and were never too preoccupied

PRISON CAR IN SIBERIA. FAMILIES OF PRISONERS AT THE WINDOWS

to lend a helping hand, or to supplement our exceedingly limited Russian. We discovered a "fourth-class guardian angel," who took us under his special protection, and was never weary of offering little kindnesses. He even wished to share with us his black bread and some curds, which we found it difficult to refuse without hurting his feelings.

None of the occupants of our car were intemperate or noisy, and, in genuine politeness of heart, these Russian serfs of a few years back could have given many points to the two selfish, unmannerly, foul-mouthed Frenchmen of the party, who represent the nation erroneously supposed to have a superfluous supply of politeness and good manners.

XII

THE HOLY SEA

ON the fifth day after leaving Stretinsk we were aroused at two o'clock in the morning, just as the faint dawn began to streak the east, by a guard going through the train and crying out: "Misovaia, all change." Eagerly we rubbed our sleepy eyes open, for now we had reached Lake Baikal, and we fondly hoped that the most trying part of our journey was over. But again we were doomed to disappointment, for we waited and waited and waited, and still there was no sign of departure to the other side of the lake. The sun rose, ascended high in the heavens, reached the zenith, and began to decline, and still we sat about miserably upon our baggage, not daring to go far lest at any moment the "ice-breaker" from the other side of the lake might come to transfer us across.

The station-master was eagerly importuned, but knew nothing. No official had the slightest information. Our Russian fellow-travellers only shrugged their shoulders when asked when we should start. At length patience ceased to be a virtue, and a telegram was prepared for the governor of Irkutsk, asking why a trainload of inoffensive passengers of all nationalities were kept in anxious suspense for so many hours at a miserable little station which had been eaten bare of provisions long ago.

Just as this telegram was about to go came word that the "ice-breaker" on the other side was waiting for the minister of justice, and that when he got ready to start the boat would come. This proved to be the case, and at last, fourteen hours after we had reached Misovaia, the great steamer which should carry us across was seen approaching. Then it took hours more to unload and reload her, and finally, in the middle of the night, twenty-two hours after we should have started, we got under way for the other side of the lake.

I dwell for a moment on this irritating experience, for it represents one phase of Russian life which is very foreign to life in free, constitutional countries. For twenty-two hours three hundred people were made to wait upon the convenience of one official. Hungry passengers were allowed to half starve; sick and crying babies, who had with difficulty survived a week of hard journeying amid unnamable privations; weary mothers, trying to comfort the babies; business and professional men with important engagements to keep, and whose connections with other roads were seriously imperilled by this delay, all must wait and suffer while this one official finished his morning nap, or ate an elaborate dinner, or otherwise made himself ready to start for the other side of Lake Baikal.

Had this happened in any free country,

THE ELEPHANT'S NOSE, LAKE BAIKAL

such a howl would have arisen as would have insured the decapitation of this minister of justice (minister of justice indeed! minister of injustice rather) at the first possible election.

How the *New York Tribune* or the *London Times* would have blazed with indignant protests had such an affair taken place in America or Great Britain! But in Siberia such an event is quite a matter of course. We were still more surprised by the utter indifference and nonchalance of our Russian fellow-passengers. There was nothing to be said and nothing to be done. "Whatever is, is right," especially if it is done by one high in authority, seemed to be their motto.

The contrast between the foreign passengers and the Russians during this exasperating delay was marked and amusing. The former champed at the bit, and fretted and fumed and (it is feared) swore in their different tongues. They would gather together in little groups and discuss the situation. They haunted the telegraph office. They interviewed every promising official. They tried to get another steamer. They exhausted every possibility of progress.

The Russians placidly sipped their tumblers of tea, gathered about the steaming samovar, quietly curled up on their luggage, and went to sleep as though they had not an irritating care in the world. Why should they fret? The official had all power. They had no appeal. No protest was open to them. No *Times* would thunder in St. Petersburg. No *Tribune* would voice their complaints in Moscow. "Let it go; life is too short to wear it out with such worries," seemed to be their mental attitude.

The irritations of those hours of waiting, however, could not destroy the beauty of the placid lake, the Holy Sea (Bai-Kal), on whose shores we were waiting.

Baikal is one of the great lakes of the world. It covers thirteen thousand five hundred square miles of surface. Only

Lakes Superior, Michigan, and Huron, of our own great lakes, surpass it in size. It is half as large again as Lake Erie, and twice as large as Lake Ontario, and being long and narrow, and surrounded by high, precipitous hills, is more picturesque than any of them. Its waters are crystal clear, and, at a depth of forty feet, every pebble upon the bottom can be seen. Drop a stone into its clear waters from the steamer's deck, and you can see it fluttering and zigzagging its way down as a feather falls through the air. The water is as cold, too, as it is clear, as a delicious bath that I took testified. Even on the last days of June the water cannot be far above the freezing point.

The peasants who live on the shores of the lake regard it with almost superstitious reverence and love. There is in their estimation no lake in all the world like the Baikal. Its rock-bound shores, its wooded steeps, green with larch, and fir, and pine, and birch; its limestone and marbles glistening in the sun; its shores strewn with bright pebbles, washed smooth and round by countless centuries of the lapping waves, all combine to make the Baikal worthy of the love and reverence with which the true Siberian regards it.

We found it in a peaceful mood, but it is not always so by any means. "No man has ever said his prayers until he has ventured on Lake Baikal," is a common saying of the peasants on the shore. Lying, as it does, in a comparatively narrow valley, between the mountain ranges, the terrible gales which have their homes between the mountain peaks are often let loose with but scant warning. Then woe betide the little fisherman's craft that is caught far from shore, and even the passengers on the stout steamer may well tremble.

To unite the two shores with a strong, substantial ferry-boat which can run throughout the winter months, the railway has had constructed in England and brought out piecemeal the huge "ice-

THE ICE-BREAKER ON LAKE BAIKAL

breaker," so called. Several of the parts of this great craft were lost, strayed, or stolen on the way, and infinite difficulty was experienced in putting it together again when it reached the lake. Even now some tell me that it is no great success as an ice-breaker.

On a fine day, with a smooth sea, nothing could be more delightful than a little journey on the great ships. Into the hold of the steamer runs the whole train without unloading; or, rather, that is to be the way when the railway is in runnnig order. Three parallel tracks throughout the entire length of the boat afford ample room for the longest trains. Above are elegant accommodations for first-class passengers, while the steerage people dispose of themselves in the hold as best they can.

Four great stacks belch out smoke and cinders, and the engines are of the most powerful pattern. A half acre of deck room over all gives the passenger unrivalled opportunities of viewing the magnificent lake, whose southern end he is about to traverse.

A fine breakwater of wooden piles, nearly half a mile long, provides a safe little harbor on the eastern side of the lake, while another, not quite so large, affords a refuge from the winds on the other side. Between these harbors the ice-breaker plies back and forth, making the distance of about fifty miles in four hours.

As we approach the western shore, the scenery grows more lovely than on the eastern side. The crystal lake, fifty feet deep at the shore, dashes in places against precipitous cliffs; the banks are clad in freshest grass, dotted with poppies and lilies and blue honeysuckles, while a little village nestles in a cove beneath a frowning cliff. Rarely have I seen in all my travels a scene more lovely than is presented as the ice-breaker makes her way cautiously into the little artificial harbor of Listvinitschnoie, at the southwestern end of the Holy Sea.

In the early morning we could see, as we thought, the seals disporting themselves in the icy waters, for it is a singular fact that in Lake Baikal alone, of all the fresh-water lakes in the world, are seals found. It is thought by some that in prehistoric times they made their way up the River Yenisei, that flows from the lake into the Arctic Ocean, two thousand miles away. At any rate, they have made themselves very much at home in the Baikal, and have increased and multiplied and replenished the lake.

The determination of the Russian government to have through steam communication between the oceans is shown by the fact that while such sums have been spent for an ice-breaker and for harbors to make the link across the Baikal possible, the railway is also being pushed around the southern end of the lake, only a short distance away. Both of these routes will be needed, I understand, to facilitate the mobilization of her armies—the supreme object which Russia has in mind in the construction of this great iron road.

At the village of Listvinitschnoie, on the western side of the lake, we had another exasperating delay of eight hours, while passengers and baggage were being transferred to the other train, a transfer which would have been made in America in half an hour. For half a day we sat about on our baggage in the broiling sun, surrounded by filthy Moujiks, while the leisurely engines puffed back and forth, and the still more leisurely porters loaded the train. At length all was ready, and we steamed in a leisurely fashion to Irkutsk, the great city of Central Siberia.

Here we arrived after being six days, within a few hours, on the road since leaving Stretinsk. In those six days we had travelled less than eight hundred miles, and less than six miles an hour, including the stops. We had endured many discomforts, some of them entirely unnecessary had a little system and forethought been used. Even in the

THE SHORES OF LAKE BAIKAL

present incomplete state of the equipment of the road the journey ought to be made in two days, and decent cars might be provided for decent people.

I must say that this experience somewhat weakened my faith in the boasted energy and enterprise of Russian railway construction. How a road that is so slow and halting in peace can be great in war, I do not see. How a line that cannot carry a few hundred passengers and their luggage without interminable delays can be used to mobilize a hundred thousand soldiers at short notice is difficult to comprehend. It is fortunate for Russia that she has a slow, stupid opponent on the other side of the world to contend with, an opponent that not only has no slow railroads, but no railroads at all, of her own.

As it is, the Trans-Baikal railway, poor as it is, will undoubtedly prove a very important instrument in forcing beneficent Russian fetters upon poor, decrepit, worn-out China.

XIII

A SIBERIAN CITY OF CHURCHES

AS the feeble and halting train that brought us from Lake Baikal, following the meanderings of the swift Angara River, drew near to Irkutsk, the capital of Central Siberia, we received the impression of a small city dominated by a score of great churches. And this impression was not dissipated when we crossed the Angara on the bridge of boats that links the city with the railway on the other side and found ourselves within the city precincts.

You cannot look up without seeing a beautiful dome or spire. In the waters of the Angara the great cathedral is reflected, flanked by two other churches which in any other city of its size would be considered marvels. The principal street of the city seems to lead up to and terminate in another lofty and imposing church, while every section of the town has its own ecclesiastical buildings of lesser magnitude. Brooklyn must look out for its laurels as the City of Churches when it comes in competition with Irkutsk.

The interior of the churches is quite equal to their exterior, and they blaze with sacred pictures and icons, framed and matted in gold, so that the chief impression one gets is of walls of solid gold, chased and fretted and highly ornamented, with the face of Christ or the Virgin or some Oriental saint peering out between the shining plates.

The Russians are evidently an extremely devout and religious people. This is evidenced by their churches, which are always, both metaphorically and literally, higher than the chimneys of their factories. In the small towns along the line of the railway the church is always the one conspicuous object. The village may be, it probably is, built entirely of logs, without a single frame or brick house within its border. But the church, with its glittering dome and blue and green towers, always lifts its Greek cross skyward, and reminds the poor Moujiks that there is another world above the flat plains of Siberia.

The unabashed devoutness of the peasants tells the same story as their churches. Our fellow-travellers in the fourth-class cars from Stretinsk never omitted their morning devotions, their ablutions being an entirely secondary matter. The poorest boatman would always cross himself and bless his crust of black bread before he ate it. The soldiers with whom we journeyed for hundreds of versts, undismayed by the presence of their comrades, would every morning face the rising sun, and though they could secure no possible privacy, would, with genuine devoutness, pray for ten minutes at a time, standing bareheaded and reverential in the crowded

BURIATS FROM LAKE BAIKAL, WITH SEAL FROM THE LAKE

railway car in which they were travelling. Such genuine devotion speaks well for the church which fosters it, and for the country that is peopled and defended by faithful devotees, for no nation without a religion which took strong hold of the hearts of the common people ever greatly influenced the destinies of the world.

Distance lends considerable enchantment to Irkutsk, and nearer approach does not bear out the remoter promise of its beautiful churches. The streets are poorly paved and extremely dusty, and the wooden sidewalks are frequently man-traps for the unwary passenger. Still it is no mean city to be found in the heart of Siberia, three thousand five hundred miles from the nearest European capital.

Irkutsk has had time enough to grow, for it was founded thirty years after the *Mayflower* sailed into Plymouth harbor, and is about ready to celebrate its quarter millennial. Founded originally as a military post to withstand the attacks of wandering Tartars, its naturally advantageous situation has made it the metropolis of Siberia. It is now the seat of government of the province of Eastern Siberia, and contains something over fifty thousand inhabitants. It contains twenty-five Greek churches, substantial Catholic and Lutheran churches, and two synagogues. There are no less than forty schools, among which are seminaries for girls, military and technical schools of various kinds, so that the young idea of Irkutsk has no excuse for not learning how to shoot.

The city has lost something of its ancient importance since the region of the Amour has been erected into a separate province, independent of Irkutsk, and since the rise of Vladivostock as a point of departure on the Pacific for Odessa and America. Nevertheless, it will always be an important centre, owing to its situation near the head-waters of the Angara, which starts from Lake Baikal, forty miles away, and continues its

AN IRKUTSK DROSKY

majestic and ever-widening way for three thousand miles until, as the Yenisei, the fourth greatest river in the world, it empties into the Arctic Ocean. It is a centre for the shipment of gold, and in a single year no less than twenty-one thousand pounds weight avoirdupois of the precious metal passed through Irkutsk on its way to St. Petersburg.

It may make the geography of Siberia a little more plain if I here explain that Siberia, for governmental purposes, is divided into six provinces: First, starting from the Ural Mountains, Western Siberia, that borders on Europe, and in a rough way embraces the basin of the Obi. Next, Eastern Siberia, which stretches a thousand miles farther east and embraces the basin of the Yenisei and its tributaries. In this province Irkutsk is situated. Then comes the Trans-Baikal, the province that lies east of the beautiful lake of that name, and through which we made our slow and toilsome way as described in the last chapter. Then, east of that, the great province of the Amour, embracing the basin of the Amour and Ussuri. Last of all, bordering on the Pacific Ocean, and including the convict island of Saghalien, is the province called in Russian Primorskaia. The province of the Steppes lies south of Western Siberia, but with that our new way around the old world has nothing to do, for the railway barely grazes its edge, and does not fairly enter its territory. During the twenty-nine days of rail and river travel which I have already described, we had traversed in succession Primorskaia, Amourskaia, and Zabaikalskaia and had reached the eastern edge of Eastern Siberia.

During these twenty-nine days we had travelled nearly twenty-seven hundred miles, and, between us and Moscow, there still lay more than thirty-three hundred miles to be traversed. But the peculiar hardships of the journey were nearly over, for a *train de luxe* starts from Irkutsk for Moscow every Friday afternoon, and we were just in season to

GENERAL VIEW OF IRKUTSK

get our tickets, and secure our berths, and replenish our seedy wardrobes, and take passage on the most famous train in all the world, which, without a single change of cars, runs a distance considerably greater than from Boston to San Francisco.

We were happy enough to board this train, as can be imagined, after the bare boards and filth of the uncertain train which had carried us from Stretinsk, and were in a fit frame of mind to enjoy its luxuries and to overlook its defects. Nor were we sorry to leave Irkutsk, for though it is a remarkable place to be found so far from the confines of civilization, yet it has few attractions to long beguile the traveller; and the stifling dust of the streets, which never seem to be watered, obscure the attractions it does possess.

Shortly before our arrival news had come of the serious disturbances in China, and troops were being enrolled and equipped and hurried to the eastern coast in feverish haste. The whole square in front of the Hôtel de Russie was filled with peasants in their red blouses and top-boots, each with his tea-kettle and little bag of poor belongings. These peasants were to be shipped off that day as food for Chinese powder, if so be the Celestials have any powder left by the time these Moujiks reach Blagavestchensk or, later, Port Arthur. The whole city was full of rumors of war, and, though we could not read the Russian papers, or know the extent of the trouble, the air seemed vibrant with battle.

We were told that the train we had taken over the Trans-Baikal road would be the last passenger train that would be allowed to go through for six weeks. How we blessed our fortunate stars that we were not stranded in Stretinsk for six times seven days! We were told also that between Omsk and Irkutsk were twelve trains loaded with soldiers, bound for Manchuria, and that no private telegram to Vladivostock could be delivered within a week, so hot were the wires with government despatches.

Some of these reports were true and some were exaggerated, but evidently China's last days had come, we thought, as we boarded the *train de luxe,* and Russia intends to be in at the death.

XIV

A SIBERIAN *TRAIN DE LUXE*

THE Paris Exposition has made famous the Siberian *train de luxe*, with its moving panorama, its terminal stations at St. Petersburg and Peking, and its dinners at seven francs per head. The newspaper correspondent, too, who has seen it only in his mind's eye, as he sat at his own cosey fireside rehashing second-hand descriptions of its magnificence, has done his share to advertise it, until the wondering world has an idea that it is a veritable Waldorf-Astoria on wheels, before which all Empire State expresses on the American continent must hide their diminished heads. We read of library-cars and bath-cars, gymnasium-cars where one can make a century run on a stationary bicycle, elegant

ONE OF THE BEAUTIFUL CHURCHES OF IRKUTSK

dinners, barber-shops where passengers receive a free shave every morning, pianos, and other luxuries too numerous to mention.

As a matter of fact, the Siberian *train de luxe,* at least as it started from Irkutsk on the 29th day of June, in the year of our Lord 1900, was a rather shabby vestibuled train of three sleepers, a diner, and a baggage-car. It was luxurious, indeed, compared with the fourth-class emigrant train on which we had been journeying, but it is still many degrees behind the best American trains. It should be remarked, however, that the best cars had been sent to Paris for the Exposition, and it is doubtless true that the train we took is somewhat below the average of the Siberian *trains de luxe.* It is well worth describing, however, as an important link in the great chain that connects the Atlantic and the Pacific in this new way around the world.

First, and very naturally, comes the engine of Russian make, with a great flaring smoke-stack for burning wood. Next to it is the baggage-car with one end fitted with a bath-tub, where dusty travellers can enjoy a refreshing bath on payment of a rouble and a half (seventy-five cents), a very high price to be sure compared with the rate charged for transportation; but then the scarcest and dearest thing throughout Siberia we found to be cold water, either for drinking or bathing purposes.

Following the baggage-car was the dining-car, which was also divided into two sections, half being occupied with small tables that seat two and four each, and the other half fitted with easy-chairs for the smokers and provided with ample windows for observation. In the smoking-room was a small library of, perhaps, fifty books in the Russian, French, and German languages, but nothing in English, except two or three discarded fifth-rate novels in paper covers, evidently contributed by previous travellers.

In the dining car, two *table d'hôte*

meals were provided: a lunch of two courses at one o'clock, which cost one rouble, and a dinner at six of four courses, for a rouble and a quarter. At other times of day passengers could order what they pleased *à la carte* from a limited bill of fare.

Behind the dining-car came a second-class sleeping-car, then a first-class sleeper, and, last of all, another second-class sleeper.

These cars are all divided into compartments, or little staterooms, holding two or four people each, with a window, a table, and a wide and very comfortable berth for each person. The cars are all handsomely carpeted and upholstered in blue plush, covered, for the sake of protection, with red-striped denim. In each stateroom were convenient racks and hooks for the disposal of clothes and baggage, and in every way one could make himself as comfortable and have almost as much room as in an ocean steamer's cabin.

For two friends travelling together, or for a small family of three, as in my own case, this arrangement is far superior to America's more promiscuous and public Pullman sleeping-car. For individuals travelling alone, the Pullman plan has some decided advantages, if one does not care to be cooped up in narrow quarters for eight or nine days with any chance stranger who may be assigned to the same stateroom. The Pullman cars, too, economize space, for while from forty to forty-eight people can sleep in a Pullman car when crowded, only eighteen can find "lying-down room," in a Siberian sleeper.

One curiosity of this particular *train de luxe* was that the first and second class cars were precisely alike in every particular. In amount of room, fittings, upholstery, and comforts of every description, there was not the slightest difference between the three sleeping-cars that composed our train, while the difference of price was nearly forty roubles in favor of the second-class. One would

A PART OF OUR *TRAIN DE LUXE*

think that, these being the facts, the second-class cars would be overcrowded and the first-class would be empty. Such was not the case on this train at least, for I found every cabin in the first-class taken, and was able to get a large four-berth stateroom in one of the second-class cars for little more than I would pay for two berths in the first-class car.

At the most, the fares in Siberia are remarkably cheap. For the whole stateroom I paid less than one hundred and twenty dollars from Irkutsk to Moscow, a distance of three thousand three hundred and fifty miles; this included four fares and the supplementary price of the *train de luxe.* For the same accommodations in a Pullman car across the American continent (a shorter distance) I should have paid at least five hundred dollars. The full first-class fare in the *train de luxe* is about fifty dollars, while the second-class fares are less than thirty dollars. By the ordinary trains a considerable saving is effected over these very cheap rates.

Each car has its porter, who can usually speak German or French in addition to Russian, but the specimens that we saw were very lazy and inefficient fellows compared with the deft and obliging "George" who attends to one's wants in a Pullman car. Our porters on the *train de luxe* never thought of sweeping or dusting the staterooms, and seemed to think it an imposition to be asked to thoroughly make up the berths for day travel, preferring to leave the beds made up by day that they might have less trouble at night. All day long the lazy porters would loll about on their seats in the middle of the car, doing as little as possible, and apparently begrudging that little. Twice a week the bed linen is changed. Ample washrooms, with a single roller towel, for all guests who do not furnish their own, complete the equipment of the Siberian sleeper.

My excuse for this somewhat minute description of these accommodations is that it will be welcomed by future travellers who have to spend at least eight and one-half days in their little room on wheels.

It must not be thought that for all its name and fame the "Siberian Special" is a lightning express. In speed it would be outstripped, during all the early part of the journey, by any "huckleberry train" in America. Fourteen miles an hour, including stops, was all it pretended to make during the first five days from Irkutsk. The road-bed is new, the rails are light, and as a consequence the train plods along at a pace which an able-bodied cowboy on his bronco would outstrip. In fact I saw, one evening, a race between our famous train and a Siberian "cow-puncher," in which, for two miles or more, the horseman more than held his own. Then he pulled up with a careless wave of his hat, as though he did not consider it worth while to race any longer with so slow a rival.

The stops at the stations, too, seem most unnecessarily protracted for such a train, which might be supposed to have the right of way. Never less than ten minutes, and more often fifteen or twenty, our patient locomotive would stand panting at every little platform in the wilds before the station-master would ring his welcome bell, to tell it that it might shriek its farewell and proceed on its way.

For a number of miles after leaving Irkutsk, the railway follows the windings of the swift-flowing Angara, and then, leaving the river, strikes off across the interminable wooded prairies that line so much of the track to Moscow. The weather was exceedingly hot during the first four or five days of the journey, the thermometers in the cars frequently registering over ninety degrees Fahrenheit. If one takes the Siberian route to avoid the heat of the Red Sea in summer, he will not gain much, though, to be sure, the cool nights make the Siberian journey more tolerable. Moreover, the heat, much of the time, was of the damp, "muggy" variety so trying to bear.

A VIEW ON THE ANGARA, NEAR IRKUTSK

A plague of flies, which would have made even Pharaoh quail, added much to the minor discomforts of the trip, though they did not afflict the passengers within the cars as sorely as when they ventured out upon the platform. Here they were, indeed, intolerable, and set every pair of hands revolving about a tormented head like a windmill in a gale. The most obnoxious insect was a little black gnat, much like the midge of the Adirondacks, which in clouds filled the air, and filled as well eyes, and nose, and ears of any venturesome traveller that left the comparative refuge of his car. The natives have learned to protect themselves against these pests by wearing veils of black mosquito netting day and night. For fully fifteen hundred miles almost every man, woman, and child we met was thus protected, and it gave a peculiarly solemn and lugubrious aspect to the little villages and stations to see every one peering at you through a mask of black netting. In this funereal attire every Siberian looked as though he had lost every friend in the world, or was about to attend his own obsequies. Trainmen working on the track, flagmen signalling the engineer, women with bottles of milk and loaves of bread, station-masters in bright official uniforms, little tow-headed children with bare feet, all had gone into mourning because of the midges.

At every station were country hucksters at a little stand, ready to sell the few products of the farms to the hungry travellers. They did not do much business, naturally, with the passengers of the *train de luxe,* who had their own dining-car, but with the regular trains or the frequent special emigrant or military trains they had a rushing trade. As we approached the European frontier, prices of these country comestibles fell, and boiled eggs were to be bought for half a cent each, and a very considerable loaf of bread for two cents and a half, while milk, fried fish, chunks of roast meat, and kvass, or Russian pop-beer, were proportionately cheap.

A TRAINMAN WHO HAS TAKEN THE VEIL

At almost every siding we passed a long train laden with emigrants or soldiers on their way to China, for the Siberian reserves had been called out. The latter were stalwart, rough, honest-looking fellows, who evidently had left their little farms at a moment's notice to serve their country. They seemed to possess no uniforms or equipments, though, to be sure, these may have been stored in the close box-cars that accompanied each train. They wore their peasants' red blouses and loose breeches. Many of them were barefooted, and some of them were hatless and ragged, but, before they meet the Chinese, these raw recruits will doubtless be licked into shape by the drill-masters, uniformed and armed, and made into very presentable soldiers. They certainly have the requisite brawn, if not the brain. I was surprised to see that there seemed to be few young men among them, scarcely any under thirty, while most looked to be over forty.

The emigrants were weighted with more

impedimenta than the soldiers, but even their household belongings were pitifully scant. Every day we passed half a dozen long trainloads of men, women, and children on their way to the new Eldorado on the banks of the Amour. Most of them were too poor to afford even the hard comfort of a third or fourth class car, and they were pigged in together in ordinary box freight-cars, which were marked on the outside to carry "twelve horses or twenty-four men." Still they did not seem discontented, much less hopeless, but at every little station jumped from their stifling boxes and raced for the hot-water tank, which a paternal government always kept steaming, free of charge. Here they filled their tin pails and made their inevitable "chi" (tea), on which, supplemented by liberal portions of black bread, they chiefly subsist.

Sometimes we indulged in the luxury of giving the little emigrating urchins, bound for a far-off land, a few kopeks, when the gratitude of the parents was

SIBERIAN RAILWAY NEAR THE ASIATIC BORDER

touching to see. Rushing from the car, they would sometimes seize our hands and kiss them in the courtly old Russian style, as though we had bestowed a king's ransom on their offspring, and then as the trains parted, going east and west, would tumble back into their fifth-class boxes, while we would return to our compartments feeling as we reached them that he would be a thankless man indeed who, with such contrasts in view, should ever grumble again at the minor discomforts of a *train de luxe.*

XV

OUT OF ASIA INTO EUROPE

THE long journey in the *train de luxe* from Irkutsk to St. Petersburg, especially the Siberian part of the journey, does not afford a great variety of scenery or incident. But, if somewhat monotonous, it is a pleasing and gentle monotony. There is not a mile of desert land in all the five thousand versts between Lake Baikal and the Atlantic, or, for that matter, between the Pacific and the Atlantic, by the way we have travelled.

There are no long stretches of sagebrush plain, such as Wyoming, Utah, and Nevada afford; no "bad lands," such as are found in the Dakotas; no interminable stretches of wearisome, treeless prairies, such as the Canadian Pacific traverses in Manitoba and Assiniboia. There are occasional wide reaches of prairie or meadow land, but trees are always in sight, and for the most part the road passes through a heavily wooded country. Birches abound everywhere, with their graceful white boles, and there are many forests of pine and poplar. The trees in the vicinity of the railway are not large, the "monarchs of the forest," if there ever were any, having been cut off at the advent of the railway.

The piles of cut wood by the railroad track impress the traveller with the boundlessness of the forests. Millions of cords of clean cut birch and pine, with scarcely a knot or a blemish, stand in neat piles waiting the insatiable engine.

Everywhere the land seems fertile, and, though but an infinitesimal part of it is cultivated, it would all reward the husbandman. Great flocks of sleek cattle roam over these natural pastures, but there is room for a hundred flocks where one roams to-day. And all this vast domain has been waiting for settlers at Europe's back door during all the centuries. Only now is the door fairly open, and the new settlers are pouring through. America, meanwhile, has been discovered, opened, and occupied; Australia has been discovered and great cities built upon its seaboard; the tropics and the poles have been explored; and all the time a great fertile domain, vaster than all Europe, and separated from Europe's swarming, overcrowded millions by no barrier of sea, and but a low mountain range, has been waiting, almost tenantless, for a people. But now Siberia has been discovered—discovered by the settler and the home-maker, as well as by the politician and the autocrat who desires a Botany Bay for his convicts—and a new and virgin empire that can afford asylum for the surplus population of two continents has been added to the world's productive domains. Undoubtedly this enormous and almost untrodden empire gives Russia a great advantage

ZLATOOST, A BORDER TOWN BETWEEN EUROPE AND ASIA

among the nations of the world. Here is a vast new continent contiguous and near to ancient Russia, a domain which cannot be seriously threatened by any enemy, a domain which in the decrepit state of China requires no fleets or armies to protect. At length Russia has awakened to a sense of its value, and is pouring into it by the ten thousand every month the peaceful settlers who will make her posession secure.

Between Irkutsk and Moscow the railway passes through a few large towns of considerable importance. Chief of these are Crasnoiarsk, about seven hundred miles from Irkutsk; Taiga, some three hundred miles still nearer Europe; Omsk, about five hundred miles further west; and Tschelabinsk, which lies near the edge of Europe, two thousand miles from our starting point at Irkutsk.

These places will be but names to most of my readers, and they will not be much the wiser if I tell them such facts as that Omsk has forty thousand inhabitants and

Tschelabinsk has fifteen thousand. All these towns, and many smaller ones that I might mention, are places of no little importance in the districts of which they are the centres, but they do not as yet cut much of a figure on the world's maps. They all have the characteristics of frontier towns, are built largely of logs, but boast from one to twenty beautiful churches, which always strike the eye and give them a character and a beauty which they would sadly lack without the shining domes and brilliant crosses pointing heavenward.

Omsk is the most important place on the line of the railway between Irkutsk and Moscow, and is the residence of the governor-general of the province of the Steppes, and of a Greek Church archbishop.

Tomsk is a larger and more important place still, but is not directly on the Trans-Siberian route, being connected by a branch line at Taiga.

The road crosses many large rivers during these two thousand miles before it reaches the European frontier, and the cost of the great iron bridges which span them must have been enormous. They seem to be built in the most substantial manner, with piers and buttresses of massive cut stone. These rivers belong to the great systems of the Yenisei and the Obi, which in their upper reaches branch off into many streams, each of which, in any other continent, would be a notable and famous stream. But Siberia is so rich in great rivers that even the names of these mighty tributaries are not of great interest to the reading public. When Siberia shall come to her own and take her place among the great populous nations of the earth, these rivers will add enormously to her wealth and commerce.

On the evening of the fifth day from Irkutsk we went to bed in Asia, and woke up on the morning of the sixth day in Europe, and in the beautifully situated town of Zlatoost. We found ourselves also in the heart of the Urals, and the contrast with the flat plains of Siberia was most refreshing.

Lovely spruce-clad mountains towered above us on every side; a rushing river, which the track follows for many miles, seemed to beckon us on with its smiling ripples that glinted and glowed in the morning sun. Great, open, park-like valleys and uplands lay between the mountains, dotted with trees and covered with the greenest of grass, studded with beautiful flowers of every hue. Altogether, for many miles, the scene was one of rare loveliness, and was worth travelling a thousand weary versts to see.

The Russians call this region of the Urals the "Switzerland of Russia," and though it lacks the grandeur of the snow-clad Alps, it has a beauty of its own which is exceedingly charming.

All day long our road wound through these charming scenes, up hill and down dale, usually keeping a rushing, sparkling river in view, until, towards the close of the long day, the great town of Oofa,

IN THE URAL MOUNTAINS

with its beautiful churches and fine railway station, came in view. We had passed the barrier of the Urals, and the day of loveliest scenery in all the long journey had come to an end.

Our deliberate train mended its pace very considerably after getting into Europe, and, on the last three days of the journey, did almost as many versts as during the first five. The engineer actually seemed able to get by some small and insignificant stations without stopping, and where we did stop it was often for less than a full half-hour. We averaged fully twenty miles an hour after crossing the Urals, which seemed like lightning speed after our slow and toilsome way in Siberia.

On the seventh day of our journey from Irkutsk, the Urals had faded into the distance, and the flat plains of European Russia were substituted for the flat plains of Asiatic Russia. The country on both sides of the mountains looks much the same, except that the European plains are better cultivated and less wooded. They have evidently known the plowshare for many a hundred years, and in early July were smiling with luxuriant crops of wheat, and barley, and rye. Greater Russia plainly is in no danger this year of a wide-spread famine.

In some places hay-making had begun, and the great unfenced fields containing hundreds of acres were alive with haymakers in gay attire. The Russian farmers live together in villages, instead of in isolated farmhouses, as do our husbandmen, and the business of cutting and making a great communal hayfield was evidently an important matter that brought together all the men, women, and children of the village. Often forty hay-wagons, each with a span of horses and five times as many people, were assembled at the edge of a field. The mowers, in long rows, would then start for the other side of the great field, nearly a mile away, cutting a wide swath as they went, while fifty Russian Maud

Mullers, in red and blue bodices, raked the meadows sweet with hay, which yesterday had been cut. They seemed merry and light-hearted about their task, and we could hear snatches of song and bursts of laughter, and many a bright bandanna was waved at the flying train.

The small villages did not strike me, as I saw them from the train, as being superior to the Siberian towns. They are built largely of logs, or of mud and wattle, and have low-browed thatch roofs, often much out of repair. But every village, however small and shabby, has a beautiful church, which contrasts strangely with the one-story thatch-roof huts that gather around it, as if a brood of very diminutive chickens were clustering around a big brooding mother hen.

The large towns which we passed, as we drew near to Moscow, were distinctly better built than in Siberia. Large blocks of brick and stone, wide streets, and a general air of thrift and prosperity seemed to pervade them.

Samara is a city nobly situated on the banks of the Volga, and has really a regal appearance on her crown of hills, with her great cathedral and noble churches dominating all.

The railway crosses the Volga, a few miles after leaving Samara, by a splendid iron bridge, nearly a mile long. Even here, so far from its source, this greatest river of Europe is a majestic, swift-flowing stream, bearing upon its ample bosom great rafts from the pine forests near its source, and many great steamers, which carry passengers to the famous fair at Nijni-Novgorod, and other towns along its banks. Most, if not all of these steamers, use petroleum for fuel, brought from the great wells of Batoum, and huge tanks for the storage of oil begin to form conspicuous features of the landscape near every large town.

Indeed, our own locomotive, instead of gorging itself with birchwood, as its predecessors which drew the *train de luxe* had done, took long deep draughts of

THE KREMLIN IN WINTER

crude kerosene oil, which, it must be said, did not improve its breath. Some of the engines I noticed at this end of the line were Baldwin locomotives, a make which I had not seen since we had left Khabaroffsk, on the further side of far Siberia.

On the eighth and last day of our journey on the "Siberian Special," as our train was called in railroad nomenclature, we passed through the large and commanding town of Toula, a great railroad centre, and also famous for the splendor of its churches. The stations we passed on this day were finer than any we had yet seen, and were often embowered in trees until they looked like châteaus in lovely green parks. Peasant women stood on every platform with heaps of little wild strawberries, which they sold at a ridiculously cheap price, about two cents a quart. Others had the first red cherries of the season, and others yellow apples that had been kept in water for nearly a year, and that tasted as if baked. Other venders of delicious, sparkling koumyss, milk, Russian kvass, and other more harmful beverages, also appeared upon the scene at many stations.

The devout character of the Russians was shown by the fact that every third-class waiting-room had its shrine, with beautifully framed pictures of Christ and the Madonna and some of the Eastern saints. Before these icons often burned ceremonial candles, and smaller candles were to be had by devotees for two or three kopeks apiece. Rapt travellers, with uplifted eyes, were often to be seen crossing themselves before these icons.

We still passed frequent trains of Siberian settlers, and others of brawny soldiers, "bearded like the pard," and packed into their fifth and sixth-class cars like the proverbial sardines in their boxes. Occasionally a train of prisoners looked out through their iron gratings at us, seeming to cast a murky shadow on the pleasant landscape, as we thought of the dreary years of exile before them. We saw no political prisoners, however, and even the criminals are allowed to take their wives and children with them. These latter often clung to the iron bars with their faces close against them, like so many infant Charlotte Cordays. It must be said that these prisoners, like others whom we had seen, fared a good deal better in their railway accommodations than the emigrants or the soldiers.

More and more cultivated and attractive the beautiful country appeared as we approached Moscow, giving evidence in every smooth and fertile field of five hundred years of tillage, until at last, promptly on schedule time, 197 hours after leaving Irkutsk, the Siberian Special rolled into the beautiful station of sacred Moscow, and our long journey across all the Russias was practically finished. It is a matter of thirteen hours further to St. Petersburg, and two days more by the fastest train to London. Thence six or seven days more by one of the Atlantic liners to New York, and our journey by this new way around an old world was completed.

SOME FACTS AND FIGURES

As a multitude of friends and other correspondents have asked me to give full particulars concerning the time and cost of this journey, I will add a few lines to answer these natural inquiries.

I was told by the most sanguine, before starting, that the journey from Vladivostock to Moscow could be made in twenty-two days. I do not think that it ever was, or can be, made in that length of time, until the Trans-Baikal Railway is greatly improved, and the Amour and Shilka have a channel dredged through their shallows; and that will be many a long day yet.

As a matter of fact, the journey took us thirty-eight days, and, allowing for probable detentions by rail and river, it cannot often be made in less time. If it had not been for the kind personal interest

A SHRINE AT A RAILWAY STATION

of the governor of Vladivostock, we should have been much longer on the way than we were. I can well imagine that the trip might take three or four months instead of as many weeks. The exact time of the different divisions of our journey was as follows:

	Days
Vladivostock to Khabaroffsk (rail)	1¼
Khabaroffsk to Blagavestchensk (boat) . . .	5½
Waited for steamer at Blagavestchensk . . .	2
Blagavestchensk to Stretinsk (boat)	12½
Waited for train at Stretinsk	1½
Stretinsk to Irkutsk (rail)	6
Waited for train at Irkutsk	1
Irkutsk to Moscow (rail)	8½
Total	38¼

If we had found high water in the Amour, we might have saved three days on the river journey, and if the bridge on the Trans-Baikal road had not burned down so inopportunely, we might have saved two days more, but it is difficult to see how more than five or six days could have been saved by the most favoring combination of circumstances.

I also heard many absurd rumors of the cheapness of the journey across Siberia. One Russian official of note told me that the whole cost from Vladivostock to Moscow, first-class, meals included, would be one hundred and twenty roubles. I will give the fares as they actually were paid, leaving out the odd kopeks:

	Roubles
Vladivostock to Khabaroffsk	17
Khabaroffsk to Blagavestchensk	23
Blagavestchensk to Stretinsk	29
Stretinsk to Irkutsk	18
Irkutsk to Moscow	98
Total for fares (first-class)	183

To this must be added at least three roubles a day for living expenses along the road, even by those economically inclined who drink no strong liquors, which will add to the expense of a thirty-eight days' journey one hundred and fourteen roubles, making the journey cost two hundred and ninety-six roubles, or, in round numbers, about one hundred and fifty dollars in gold—not expensive, to be

RUSSIAN TYPES

sure, for a six-thousand-mile trip, but not so cheap as is usually thought.

Those who are willing to travel in second-class cars and steamer accommodations can save nearly one hundred roubles, or about fifty dollars, on the whole journey. The second-class cars are entirely comfortable, and very little poorer than the first-class, but the second cabin on the steamers is not to be recommended.

I have very often been asked whether I would advise others to take the journey. That all depends upon their circumstances and travel proclivities.

For good sailors who do not mind a long ocean voyage, the trip through the Suez Canal is certainly easier and more restful. For bad sailors and those who fear the sea for any reason—and their name is legion—the Trans-Siberian route offers many compensations. Moreover, if one can command his own time, and is in no great haste to reach his journey's end at a fixed date, the Siberian journey will prove far more attractive than to those who are pressed with engagements and must make schedule time, for it is impossible to calculate exactly, or even within a fortnight, the time that may be taken. A pressing engagement in London, which I almost missed owing to the unexpected delays, detracted not a little from the comfort of the journey so far as I was concerned.

Nevertheless, in spite of the discomforts and delays, I am exceedingly glad to have taken the journey. It was an experience never to be forgotten, and furnished memory-pictures that will never fade. I should advise any one with plenty of time on his hands, who enjoys rail and river travel, and who is not afraid of the inevitable hardships of a new frontier country, to go and do likewise.

The journey I have described was made at a most fortunate time, immediately before the Chinese outbreak, which may change the face of the world, especially of the far Siberian world. A week later, or a fortnight later at the most, the journey would have been impossible for many months. But doubtless before my readers wish to take this new way round the old world the "Yellow Peril" will be overpassed, and the Trans-Siberian route will be once more open to the travelling public.

In these annals I have purposely omitted any extended reference to the commercial and political importance of the Trans-Siberian Railway, for the reading public has been flooded with speculations and facts, more or less accurate, on this subject. There seems, then, to be little need for their further exploitation. Moreover, history, during the next decade, will determine these matters much more accurately than the wisest prophet can forecast them to-day.

When the Eastern Chinese Railway, connecting with the Trans-Siberian road, is opened—if the Mongolians, who are showing such unexpected ferocity, ever allow it to be opened—it will be possible, perhaps, to go from Paris to Peking in two weeks, or, in time, even in eleven days, as Mr. A. R. Colquhoun, in his excellent book, *Overland to China*, points out.

Even then, there will be many who will desire to take the journey I have described through the lordly domain of Siberia, by rail and river from the Pacific to the Atlantic, which for vastness of extent along the same east and west parallels, for variety of primeval scenery, and for wideness of view over "this goodly frame, the earth," will never be surpassed.

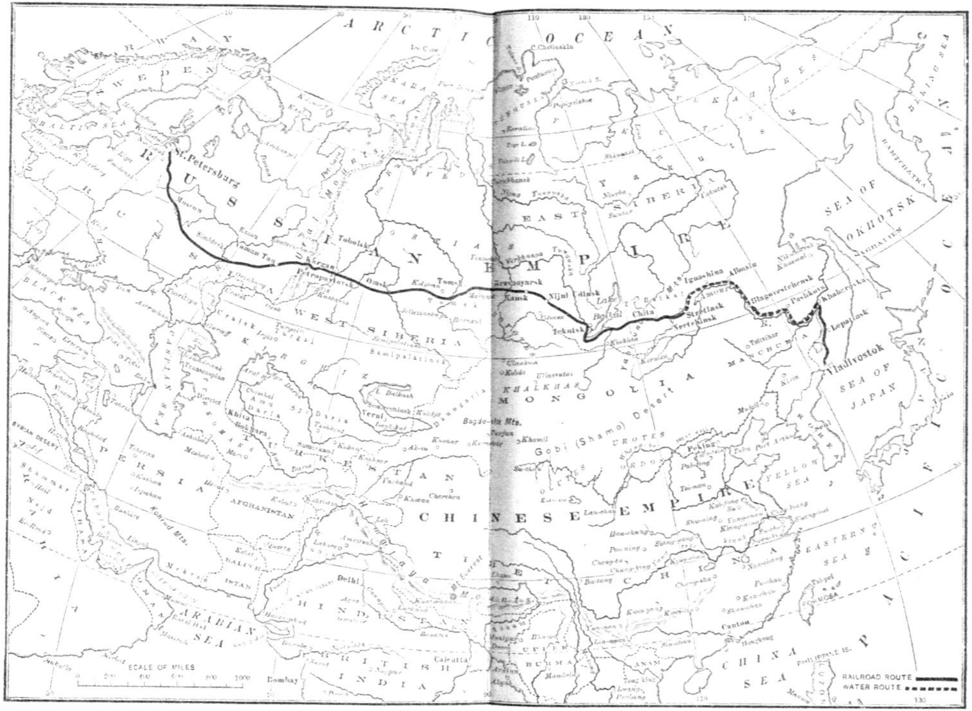

MAP SHOWING THE AUTHOR'S ROUTE FROM VLADIVOSTOCK ACROSS SIBERIA TO ST. PETERSBURG

Synopsis of the Russian Alphabet

Synopsis of the

Form.	Proper sound.	Name.	Form.	Proper sound.
А а	a *in* father	ah	Л л	ll *in* dollar
Б б	b *in* butter	bey	М м	m *in* man
В в	v *in* vein	vey	Н н	n *in* note
Г г	g *in* gone	ghey	О о	o *in* not
Д д	d *in* done	dey	П п	p in pit
Е е	ye *in* yet	yey	Р р	rr *in* hurry
Ж ж	z in azure	zhey	С с	s *in* safe
З з	z *in* zone	zey	Т т	t *in* turn
И и	i *in* sick	ee	У у	oo *in* moon
І і	i *in* iota	ee	Ф ф	f *in* fine
Й й	i *in* oil	ee s'krátkoy	Х х	ch *in* Dach
К к	k in kin	kah	Ц ц	ts *in* wits

Russian Alphabet

Name.	Form.	Proper sound.	Name.
ell	Ч ч	ch *in* **ch**arm	chey
emm	Ш ш	sh *in* **sh**ut	shah
enn	Щ щ	shtch *in* sma**sht-ch**ina	shtchah
oh	Ъ ъ	*mute*	*hard sign*
pey	Ы ы	*approximately* y *in* pit**y**	yairrwee
airr	Ь ь	*half mute*	*soft sign*
ess	Ѣ ѣ	ye *in* **ye**t	yaht
tey	Э э	e *in* met	a
oo	Ю ю	u *in* **u**niform	you
eff	Я я	ya *in* **y**ard	yah
khah	Ѳ ѳ	f *in* **f**ine	feetah
tsey	Ѵ ѵ	i *in* s**i**ck	eezhitsa

SOME NECESSARY RUSSIAN WORDS

The following list of words, of course, does not pretend to be an English-Russian vocabulary, but simply the smallest possible number of words which a traveller would need in crossing Siberia. It would be well, of course, to know many more. But it would be possible for the traveller to make this journey with the limited means of communication with his fellow-travellers, restaurant-keepers, hotel-proprietors, train-men and steam-boat officials that are here put at his disposal. As it is impossible, of course, in a volume of this character to give anything like a complete vocabulary, I have thought that a very small fraction of a loaf was better than no bread, and that this was not one of the cases when a little knowledge (of Russian) was a dangerous thing.

Cardinal Numbers

одинъ, one.
два, two.
три, three.
четыре, four.
пять, five.
шесть, six.
семь, seven.
восемь, eight.
девять, nine.
десять, ten.

English-Russian Vocabulary

Able (to be), быть въ состояніи.
accident, несчастіе.
act (to), дѣйствовать.
address (direction), адресъ, надпись.
adieu, прощай, прощайте.
air, воздухъ.
all, весь, вся, все.
America, Америка.
animal, животный.
apartment, комната, квартира.
apple, яблоко.
April, апрѣль.
as, какъ, такъ, такъ какъ.
ask (to), спрашивать.
August, августъ.

Bag, кошелекъ.
baggage, багажъ.
barber, цирюльникъ.
bath, ванна.
bed (bedstead), постель, кровать.
bedroom, спальня.
beefsteak, бифстексъ.
bill (account), счетъ.
boat, лодка.
book, книга.
bookseller, книгопродавецъ.
boot, сапогъ, ботинка.
box (coffer), сундукъ.
bread, хлѣбъ.
breakfast, завтракъ.

business, дѣло.
butter, масло (коровье)
button, пуговица.
buy (to), покупать, купить.

Cab, дрожки.
cabman, извощикъ.
can (I), я могу.
carriage, карета, телѣга.
carry (to), носить, нести.
chair, стулъ.
chamber, комната.
China, Китай.
church, церковь.
city, городъ.
class, классъ, разрядъ.
clean (to) чистить.
clock (what o'—is it?), который часъ?
clothing, одежда, платье.
coachman, кучеръ.
coffee, кофей.
coffee-house, кофейня.
copeck, копейка, (копѣйка).
cost (to,) стоить.
cure (to), лѣчить.

Day, день.
December, Декабрь.
dentist, зубной врачъ.
dinner, обѣдъ.
dirty, грязный.
dish, блюдо, кушаніе.
do (to), дѣлать.
doctor, врачъ, лѣкарь, докторъ.
dress, платье.

Eat (to), ѣсть, кушать.
egg, яйцо.
engage (to), уговаривать.
enough, довольно.
Europe, Европа.
express-train, курьерскій поѣздъ.

Face, лицо.
fact, дѣло.
family, семейство.
farmer, фермеръ.
February, Февраль.
first (at), сперва.
fish, рыба.
flour, мука.
flower, цвѣтъ, цвѣтокъ.
oreign, иностранный.
fork, вилка.
fowl, курица.
French, французскій.
Friday, пятница.
friend, пріятель, другъ.
fur, шуба, мѣхъ.

Girl, дѣвочка.
give (to), давать, дать.
glass (drinking), стаканъ.
glove, перчатка.
gold, золото.
good, добрый.

Half, половина, полу . . .
handkerchiefs, (носовой) платокъ.
hat, шляпа, шапка.
have (to), имѣть.
here, здѣсь.

hill, холмъ.
horse, конь, лошадь.
hot, жаркій, горячій.
hotel, гостинница.
hour, часъ.
house, домъ, жилище.
how much, сколько.
hungry (to be—), бытьголоднымъ.

Ice, ледъ.
ill (sick), больной.
important, важный.
impossible, невозможный.
inhabitant, житель, обитатель.
ink, чернила.
inn, гостинница.
instantly, немедленно.

January, январь.
journal, журналъ.
July, Іюль.
June, Іюнь.

Key, ключъ.
knife, ножъ, ножикъ.

Lady, госпожа.
lamp, лампа. лампада.
landlord, хозяинъ.
language, языкъ.
lantern, фонарь.
large, большой, широкій.
letter, письмо.
light, свѣтъ, сіяніе.
linen, бѣлье.
little (a), мало.

lock, замокъ.
lodging, квартира.
lunch, закуска.

Mail, почта.
man, человѣкъ; men, люди.
many, многіе.
map, карта.
March, Мартъ.
matter (no), все равно.
May, Май.
meat, мясо.
mend (to), исправлять.
mile, миля.
milk, молоко.
minute. минута.
Monday, понедѣльникъ.
money, деньги, монета.
morning, утро.
morrow (to), завтра.
Moscow, Москва.
mother, мать.

Name, имя.
needle, иголка.
never, никогда.
news, новость, извѣстіе.
newspaper, вѣдомость, газета.
night, ночь.
no, нѣтъ.
noon, полдень.
north, сѣверъ.
November, ноябрь.

October, октябрь.
overcoat, пальто.

Petersburg (Saint), (Санктъ) Петербургъ.
pillow, подушка.
plate, тарелка.
please (if you—), пожалуйста.
post-office, почтамтъ, почта.
potato, картофель.
promise (to), обѣщать.
pronounce (to), произносить.
province (Russ.), губернія.
purse, кошелекъ.
put (to), класть.

Quarter, четверть.
quickly, скоро.

Railroad, railway, желѣзная дорога.
railway-station, вокзалъ.
rain, дождь.
read (to), читать
refreshment, освѣженіе.
registered (letter), заказное.
repair (to), поправлять.
rest (repose), отдыхъ.
river, рѣка.
road, дорога, путь.
room, комната.
Russia, Россія.
Russian, Россіянинъ, русскій.

Salt, соль.
Saturday, суббота.
say (to), сказать, говорить.
sea, море.
September, сентябрь.
sheet (bedlinen), простыня.

Siberia, Сибирь.
silver, серебро.
soap, мыло.
soup, супъ, похлебка.
speak (to), говорить.
spoon, ложка.
stay (to), оставаться.
steamboat, steamer, пароходъ.
stocking, чулокъ, носокъ.
stove, печь, печка.
street, улица.
sufficient, достаточный.
sugar, сахаръ.
Sunday, воскресенье.
supper, ужинъ.
sweep (to), мести.

Table, столъ.
table-cloth, скатерть.
talk (to), разговаривать.
tea, чай.
thank (to), благодарить.
there is, вотъ.
thirsty, жаждущій.
Thursday, четвергъ.
ticket, билетъ.
towel, полотенце.
trunk (coffer), сундукъ.
Tuesday, вторникъ.

Vegetables, овощи.

Wagon, вагонъ, телѣга.
wait (to), ждать, дожидать.
waiter, служитель, человѣкъ.

waiting-room, (станцiонная) зала.
waken (to), разбудить.
walk (to), гулять, ходить.
want (to), нуждаться, желать.
wash (to), мыть, обмывать.
washerwoman, прачка.
washstand, рукомойникъ.
water, вода.
way, путь, дорога.
Wednesday, середа.
what, что; какой; то что.
when, когда.
wish (to), желать.
woman, жена, женщина.
wood (fuel), дрова.
word, слово
work (to), работать.
write (to), писать.

Yes, да.

THE END

CATALOGUE
OF
Popular Illustrated Books

Classified according to Prices.

New Books and New Editions are marked with an Asterisk.

Published by

S. W. PARTRIDGE & CO.,

8 & 9, Paternoster Row, London, E.C.

10s.

By FERRAR FENTON.

The Complete Bible in Modern English. Incorporating the five volumes recently issued. Cloth extra, gilt top. *Net.* *(For list of separate volumes, see page 7).*

7s.

The Holy Land in Geography and History. By Townsend McCoun, M.A. With 154 Coloured Maps and Plans Two volumes (not sold separately). Crown 8vo. Cloth boards.

5s. each.

The Scarlet Judges. By Eliza F. Pollard, Author of "A Hero King: A Romance of the Days of Alfred the Great," etc. Illustrated. Large Crown 8vo. Bevelled boards, gilt edges.

The Two Protectors: Oliver and Richard Cromwell. By Sir Richard Tangye, K.B. With extracts from unpublished manuscripts and reproductions of original portraits in the collection of the Author. Demy 8vo. 300 pages. Cloth boards, gilt top. 38 Illustrations on Art Paper.

The Two Babylons; or, The Papal Worship proved to be the Worship of Nimrod and his Wife. With 61 Illustrations from Nineveh, Babylon, Egypt, Pompeii, etc. By the late Alexander Hislop. Demy 8vo. Cloth extra.

By ROBERT LEIGHTON.

Cap'n Nat's Treasure: A Tale of Old Liverpool. By Robert Leighton, Author of "The Pilots of Pomona," etc. Eight Illustrations. Large Crown 8vo. Cloth.

By GEO. MANVILLE FENN.

Sappers and Miners; or, The Flood Beneath the Sea. By G. Manville Fenn. Large Crown 8vo. Illustrated. Cloth boards, gilt edges.

In the Mahdi's Grasp. By G. Manville Fenn. Illustrated by Lancelot Speed. Large Crown 8vo. Cloth boards, gilt edges.

In Honour's Cause; A Tale of the Days of George the First. By G. Manville Fenn. Illustrated by Lancelot Speed. Large Crown 8vo. Cloth boards, gilt edges.

Steve Young; or, The Voyage of the "Hvalross" to the Icy Seas. By G. Manville Fenn. Large Crown 8vo. Illustrated. Cloth boards, gilt edges.

By G. A. HENTY.

Cuthbert Hartington: A Tale of Two Sieges of Paris. By Geo. A. Henty. Large Crown 8vo. Six Illustrations. Cloth boards, gilt edges.

3s. 6d. each.

By SILAS K. HOCKING.

***Smoking Flax.** By Silas K. Hocking, Author of "In Spite of Fate," etc. Large Crown 8vo. Twelve Illustrations. Cloth, gilt top.

More than Money. By A. St. John Adcock, Author of "In the Image of God," etc. Twelve Illustrations. Large Crown 8vo. Cloth, gilt top.

Those Berkeley Girls. By Lillias Campbell Davidson, Author of "Into a Far Country." Twenty-four Illustrations. Large Crown 8vo. Cloth boards, gilt top.

The Awakening of Helena Thorpe. By E. Rentoul Esler, Author of "A Maid of the Manse," "The Wardlaws," etc. Large Crown 8vo. Eight Illustrations. Cloth boards, gilt top.

The Parting of the Ways. By J. L. Hornibrook, Author of "In the Bonds of Silence." Twelve Illustrations. Large Crown 8vo. Cloth boards, gilt top.

A Lion of Wessex; or, How Saxon fought Dane. By Tom Bevan, Author of "Dick Dale, the Colonial Scout." Large Crown 8vo. Eight Illustrations. Cloth boards, gilt top.

Among Hills and Valleys in Western China. By Hannah Davies. Demy 8vo. 326 pages. Fifty Illustrations from photos, and Large Map.

Jungle and Stream; or, The Adventures of Two Boys in Siam. By G. Manville Fenn, Author of "Cormorant Crag," etc. Large Crown 8vo. Illustrated. Cloth boards, gilt top.

A Hero King: A Romance of the Days of Alfred the Great. By Eliza F. Pollard, Author of "A Gentleman of England," "The White Dove of Amritzir," etc. Frontispiece. Gilt top.

The Three-Cornered House. By the Author of "Everybody's Friend." Illustrated. Large Crown 8vo. Cloth extra, gilt top.

Casque and Cowl: A Tale of the French Reformation. By F. M. Cotton Walker, Author of "The Lily and the Rose," etc. 364 pages. Four Illustrations. Large Crown 8vo. Cloth boards, gilt top.

A Boy of the First Empire. By Elbridge S. Brooks. 332 pages. Six Illustrations. Large Crown 8vo. Cloth boards, gilt top.

Dick Dale; The Colonial Scout. A Story of the Transvaal War of 1899-1900. By Tom Bevan, Author of "White Ivory and Black," etc. Large Crown 8vo. 332 pages. Illustrated. Cloth extra, gilt top.

A Lady of High Degree. By Jennie Chappell, Author of "Her Saddest Blessing," etc. Large Crown 8vo. Illustrated. Cloth extra, gilt top.

3s. 6d. each *(continued).*

A Gentleman of England. A Story of the Time of Sir Philip Sidney. By Eliza F. Pollard, Author of "The White Dove of Amritzir," "Roger the Ranger," etc. Large Crown 8vo. Cloth extra, gilt top.

Pilgrims of the Night. By Sarah Doudney. Frontispiece. Large Crown 8vo. Cloth extra, gilt top.

The Story of the Bible. Arranged in Simple Style for Young People. One Hundred Illustrations. Demy 8vo. Cloth extra. 3s. 6d. Gilt edges, bevelled boards, 4s. 6d.

Six Stories by "Pansy." Imperial 8vo. 390 pages. Fully Illustrated and Well-Bound in Cloth, with attractive Coloured Design on Cover, and Six Complete Stories in each Vol. 3s. 6d. each.

Through Fire and Storm: Stories of Adventure and Peril. By G. A. Henty, G. Manville Fenn, and John A. Higginson. Crown 8vo. 320 pages. Frontispiece. Cloth extra, gilt top.

The Skeleton Reef: A Sea Story. By Hugh St. Leger. Large Crown 8vo. Frontispiece. Cloth extra, gilt top.

The Pilgrim's Progress. By John Bunyan. Illustrated with 55 full-page and other Engravings, drawn by Frederick Barnard, J. D. Linton, W. Small, and engraved by Dalziel Brothers. Crown 4to. Cloth extra, 3s. 6d. Gilt edges, 5s.

The Scuttling of the "Kingfisher." By Alfred E. Knight. Frontispiece. Large Crown 8vo. Cloth extra, gilt top.

The Missing Million: A Tale of Adventure in Search of a Million Pounds. By E. Harcourt Burrage. Frontispiece. Large Crown 8vo. Cloth extra, gilt top.

Come, Break your Fast: Daily Meditations for a Year. By Rev. Mark Guy Pearse. 544 pages. Large Crown 8vo. Cloth extra.

2s. 6d. each.

"ROMANCE OF COLONISATION."

Special attention is requested to this well-written and up-to-date Series of Books on the development of British Colonisation from its commencement to the present day.

Crown 8vo. Frontispiece. 320 pages. Cloth extra, 2s. 6d. each.

I.—The United States of America to the Time of the Pilgrim Fathers. By G. Barnett Smith.

II.—The United States of America to the Present Day. By G. Barnett Smith.

III.—India. From the Aryan Invasion to the Great Sepoy Mutiny. By Alfred E. Knight.

IV.—Canada: Its Rise and Progress. By G. Barnett Smith.

2s. 6d. each *(continued).*

"RED MOUNTAIN" SERIES.

Crown 8vo. 320 pages. Illustrated. Handsomely Bound in cloth boards.

The Firebrands of the Caspian: The Story of a Bold Enterprise. By F. M. Holmes. Illustrated.

A Desert Scout: A Tale of Arabi's Revolt. By Wm. Johnston. Illustrated.

Cormorant Crag: A Tale of the Smuggling Days. By G. Manville Fenn. Eight Illustrations by W. Rainey, R.I.

One of the Tenth: A Tale of the Royal Hussars. By William Johnston. Six Illustrations.

Gold or Dross? By John W. Kneeshaw, Author of "Norcliffe Court," etc. Six Illustrations.

Lost in the Slave Land; or, The Mystery of the Sacred Lamp Rock. By W. M. Graydon. Six Illustrations.

Ice-Bound; or, The Anticosti Crusoes. By Edward Roper, F.R.G.S. Six Illustrations.

On Winding Waters: A Tale of Adventure and Peril. By William Murray Graydon. Six Illustrations.

Under the Sirdar's Flag. By William Johnston, Author of "Tom Graham, V.C.," etc. Six Illustrations.

Dorothy: The Coombehurst Nightingale. By E. M. Alford, Author of "Honor," etc. Frontispiece.

The Burtons of Burdale. By John W. Kneeshaw, Author of "Norcliffe Court," etc.

The Boy from Cuba: A School Story. By Walter Rhoades, Author of "Two Scapegraces," etc.

The Fighting Lads of Devon; or, In the Days of the Armada. By Wm. Murray Graydon.

A Trip to Many Lands. By W. J. Forster. Illustrated with 26 full-page pictures. 4to. Cloth gilt.

The Adventures of Mark Paton. By Charles J. Mansford, Author of "Shafts from an Eastern Quiver," etc.

The Adventures of Don Lavington. By G. Manville Fenn.

By Sea-Shore, Wood, and Moorland: Peeps at Nature. By Edward Step, Author of "Plant Life," etc.

The Crystal Hunters: A Boy's Adventures in the Higher Alps. By G. Manville Fenn.

The Eagle Cliff: A Tale of the Western Isles. By R. M. Ballantyne, Author of "Fighting the Flames," etc.

2s. 6d. each *(continued)*.

"RED MOUNTAIN" SERIES *(continued)*.

England's Navy: Stories of its Ships and its Services. With a Glance at Some Navies of the Ancient World. By F. M. Holmes, Author of "Great Works by Great Men," etc.

First in the Field: A Story of New South Wales. By G. Manville Fenn. 416 pages.

The Grand Chaco. By the same Author. 416 pages.

Great Works by Great Men: The Story of Famous Engineers and their Triumphs. By F. M. Holmes.

In Battle and Breeze: Sea Stories. By G. A. Henty, G. Manville Fenn, and E. Harcourt Burrage.

The Lady of the Forest. By L. T. Meade, Author of "Scamp and I," "Sweet Nancy," etc.

Loyal: A Story of the Mercantile Marine. By Arthur Collard.

Norcliffe Court. By John W. Kneeshaw, Author of "A Black Shadow," "From Dusk to Dawn," etc.

Roger the Ranger: A Story of Border Life among the Indians. By Eliza F. Pollard, Author of "Not Wanted," etc.

The Red Mountain of Alaska. By Willis Boyd Allen, Author of "Pine Cones," "The Northern Cross," etc.

True unto Death: A Story of Russian Life and the Crimean War. By Eliza F. Pollard, Author of "Roger the Ranger."

The Two Henriettas. By Emma Marshall, Author of "Eaglehurst Towers," etc.

Vashti Savage: The Story of a Gipsy Girl. By Sarah Tytler.

The White Dove of Amritzir: A Romance of Anglo-Indian Life. By Eliza F. Pollard.

POPULAR WORKS BY MISS CHARLOTTE MURRAY.

Coral: A Sea Waif and her Friends. 268 pages. Six Illustrations. Cloth boards.

Castleton's "Prep." Crown 8vo. 288 pages. Six Illustrations. Cloth boards.

Muriel Malone; or, From Door to Door. Illustrated. Crown 8vo. 272 pages. Cloth boards.

Through Grey to Gold. Crown 8vo. 280 pages. Six Illustrations. Cloth boards.

Wardlaugh; or, Workers Together. (*See under Home Library.* 2s.)

2s. 6d. each *(continued).*

FERRAR FENTON'S TRANSLATIONS OF THE HOLY SCRIPTURES IN MODERN ENGLISH. Each *Net.*

Vol. I.—The Five Books of Moses.

Vol. II.—The History of Israel.

Vol. III.—The Books of the Prophets.

Vol. IV.—The Psalms, Solomon, and Sacred Writers.

Vol. V.—The New Testament.

A Young Man's Mind. By J. A. Hammerton, Author of "J. M. Barrie and His Books," etc. Crown 8vo. Cloth extra, gilt top.

The Romance of The Bible. The Marvellous History of the British and Foreign Bible Society. By Charles F. Bateman, Author of "Dr. Clifford," etc. Crown 8vo. Cloth.

Crown and Empire: A Popular Account of the Lives, Public and Domestic, of Edward VII. and Queen Alexandra. With Notes on some Memorable Coronations. By Alfred E. Knight, Author of "Victoria: Her Life and Reign," etc. Large Crown 8vo. 336 pages. Bound in handsome cloth boards, and containing 28 Illustrations.

My Dogs in the Northland. By Egerton R. Young, Author of "By Canoe and Dog Train," etc. 288 pages. Many Illustrations. Crown 8vo. Cloth.

From Stage to Cross: The Record of a Rolling Stone. By F C. Vernon-Harcourt. Crown 8vo. 320 pages. Illustrated. 2s. 6d. Net.

Our Rulers: From William the Conqueror to Edward VII. By J. Alexander. Foolscap 4to. Cloth gilt. Sixty beautiful Illustrations. Attractively bound.

The Great Siberian Railway: What I saw on my Journey. By Dr. F. E. Clark. Crown 8vo. 213 pages. Sixty-five First-class Illustrations on art paper, and a Map. Handsomely bound.

Chaplains at the Front. By One of Them. Incidents in the Life of a Chaplain during the Boer War, 1899-1900. By Owen Spencer Watkins, Acting Wesleyan Chaplain to His Majesty's Forces. With 44 excellent Illustrations, printed on art paper, from photographs, etc., taken on the field, and 3 Maps. Crown 8vo. 334 pages. Handsomely bound.

With Kitchener's Army: Being a Chaplain's Experiences with the Nile Expedition, 1898. By Owen Spencer Watkins. Illustrated from photographs taken on the field. Crown 8vo. Cloth extra.

Lord Roberts of Kandahar, V.C.: The Life-Story of a Great Soldier. By Walter Jerrold. Crown 8vo. Eight Illustrations. Handsomely bound in cloth boards. 2s. 6d. Net.

2s. 6d. each *(continued)*.

Sir Redvers H. Buller, V.C.: The Story of His Life and Campaigns. By Walter Jerrold. Crown 8vo. 218 pages. With 8 Illustrations. 2s. 6d. net.

Following Jesus: A Bible Picture Book for the Young. Size 13½ by 10 inches. Contains 12 beautifully coloured Old and New Testament Scenes, with appropriate letterpress by D.J.D.

Brought to Jesus: A Bible Picture Book for Little Readers. Containing 12 New Testament Scenes, printed in colours. Size 13½ by 10 inches. Handsome coloured boards.

Bible Pictures and Stories: Old and New Testament. In one Volume. Bound in handsome cloth, with 89 full-page Illustrations by Eminent Artists.

Light for Little Footsteps; or, Bible Stories Illustrated. With beautiful coloured Cover and Frontispiece. Full of Pictures. Size 13½ by 10 inches.

Potters: their Arts and Crafts. Historical, Biographical, and Descriptive. By John C. Sparks and Walter Gandy. Crown 8vo. Copiously Illustrated. Cloth extra, 2s. 6d.; art linen, gilt edges, 3s. 6d.

The Story of Jesus. For Little Children. By Mrs. G. E. Morton. Many Illustrations. Imperial 16mo.

The Spiritual Grasp of the Epistles; or, An Epistle a-Sunday. By Rev. Charles A. Fox. Cloth Boards.

Upward and Onward. A Thought Book for the Threshold of Active Life. By S. W. Partridge. Cloth boards.

Victoria: Her Life and Reign. By Alfred E. Knight. New edition, brought up to date. Crown 8vo. 384 pages. Cloth extra, 2s. 6d.; fancy cloth, gilt edges, 3s. 6d.

2s. each.

THE HOME LIBRARY.

Crown 8vo. 320 pages. Handsome Cloth Covers. Artistically Illustrated.

***Edwin, the Boy Outlaw;** or, the Dawn of Freedom in England. A Story of the Days of Robin Hood. By J. Frederick Hodgetts, Author of "Older England," etc.

***Manco, the Peruvian Chief.** By W. H. G. Kingston. New Edition. Illustrated by Lancelot Speed.

Neta Lyall. By Flora E. Berry, Author of "In Small Corners," etc. Six Illustrations.

Robert Aske: A Story of the Reformation. By E. F. Pollard. Eight Illustrations.

John Burleigh's Sacrifice. By Mrs. Chas. Garnett. Nineteen Illustrations.

2s. each *(continued).*

THE HOME LIBRARY *(continued).*

The Lion City of Africa. By Willis Boyd Allen. Sixteen Illustrations.

Aveline's Inheritance. By Jennie Chappell, Author of "A Lady of High Degree," etc. Six Illustrations.

Around the Fire: Yule-tide Stories. By M. S. Haycraft.

Ben-Hur. By L. Wallis.

The Better Part. By Annie S. Swan.

A Bunch of Cherries. By J. W. Kirton.

A Child of Genius. By Lily Watson.

Cousin Mary. By Mrs. Oliphant, Author of "Katie Stewart."

Dorothy's Training; or, Wild-flower or Weed? By Jennie Chappell.

Fortune's Wheel. A South African Story. By Eliza F. Pollard.

Grace Ashleigh; or, His Ways are Best. By Mary D. R. Boyd.

Honor: A Nineteenth-Century Heroine. By E. M. Alford.

Her Saddest Blessing. By Jennie Chappell.

The Inca's Ransom: A Story of the Conquest of Peru. By Albert Lee, Author of "The Black Disc," etc.

John: A Tale of the Messiah. By K. Pearson Woods.

Jacques Hamon; or, Sir Philip's Private Messenger. By Mary E. Ropes.

Leaders into Unknown Lands: Being Chapters of recent Travel. By A. Montefiore Brice, F.G.S., F.R.G.S.

Lights and Shadows of Forster Square. By Rev. E. H. Sugden, M.A.

Living it Down. By Laura M. Lane.

The Last Earl Grahame. By Rev. J. M. Dryerre, LL.B., F.R.G.S., Authors of "Heroes and Heroines of the Scottish Covenanters." Six Illustrations.

More Precious than Gold. By Jennie Chappell.

The Martyr of Kolin: A Story of the Bohemian Persecution. By H. O. Ward.

Morning Dew-Drops: A Temperance Text Book. By Clara Lucas Balfour.

Mark Desborough's Vow. By Annie S. Swan.

2s. each *(continued).*

THE HOME LIBRARY *(continued).*

Mick Tracy, the Irish Scripture Reader. By the Author of "Tim Doolan, the Irish Emigrant."

Norman's Nugget. By J. Macdonald Oxley, B.A. Author of "Archie Mackenzie," etc. Six Illustrations.

Out of the Deep: A Story of the Brays of Beachtown. By E. Harcourt Burrage.

A Puritan Wooing: A Tale of the Great Awakening in New England. By Frank Samuel Child.

Petrel Darcy; or, In Honour Bound. By T. Corrie.

A Polar Eden; or, The Goal of the "Dauntless." By Charles R. Kenyon, Author of "The Young Ranchman," etc.

The Strait Gate. By Annie S. Swan.

The Spanish Maiden: A Story of Brazil. By Emma E. Hornibrook.

The Tramp Ship's Fate: The Story of a Secret Commission. By F. M. Holmes. Six Illustrations.

A Village Story. By Mrs. G. E. Morton, Author of "The Story of Jesus," etc.

Without a Thought; or Dora's Discipline. By Jennie Chappell.

Wardlaugh; or, Workers Together. By Charlotte Murray.

The Wreck of the "Providence." By Eliza F. Pollard.

Alfred the Great: The Father of the English. By Jesse Page. Crown 8vo. 288 pages. Eight Illustrations. Cloth extra.

LIBRARY OF STANDARD WORKS BY FAMOUS AUTHORS.

Crown 8vo. Bound in handsome cloth boards.

***Grimm's Fairy Tales.** Carefully chosen from the Tales collected by the Brothers Grimm. Twelve Illustrations. 344 pages.

***The Swiss Family Robinson:** Adventures on a Desert Island. Twelve Illustrations. 400 pages.

Tom Brown's School-Days. By an Old Boy. 344 pages. Twelve Illustrations.

Little Women and Good Wives. By Louisa M. Alcot. 450 pages. Six Illustrations.

The Wide, Wide World. By Susan Warner. 478 pages. Six Illustrations.

2s. each *(continued).*

LIBRARY OF STANDARD WORKS BY FAMOUS AUTHORS *(contd.)*

Danesbury House. By Mrs. Henry Wood. 332 pages. Six Illustrations.

Stepping Heavenward. By E. Prentiss. 332 pages. Six Illustrations.

John Halifax, Gentleman. By Mrs. Craik. New Edition. 540 pages.

Life and Adventures of Robinson Crusoe. By Daniel Defoe.

Naomi; or, The Last Days of Jerusalem. By Mrs. Webb.

The Pilgrim's Progress. By John Bunyan. 416 pages.

Uncle Tom's Cabin. By Harriet Beecher Stowe.

Westward Ho! By Chas. Kingsley.

Bunyan's Folk of To-day; or, The Modern Pilgrim's Progress. By Rev. J. Reid Howatt. Twenty Illustrations. Crown 8vo. Cloth extra.

Sunday Afternoons with My Scholars. By J. Attenborough. With portrait. Crown 8vo. 290 pages. Cloth gilt

Bible Light for Little Pilgrims. A Coloured Scripture Picture Roll. Contains 12 beautifully coloured Old and New Testament Scenes, with appropriate texts. Varnished cover printed in 10 colours. Mounted on Roller for hanging.

Twilight Whispers. For Devotional Moments. By. J. O. Keen. Crown 8vo. 256 pages, with Portrait of the Author. Cloth boards.

Platform, Pulpit, and Desk; or, Tools for Workers. Being 148 Outline Addresses on all Phases of the Temperance Movement for all Ages and Classes. By W. N. Edwards, F.C.S. With an Introduction by Canon Barker. Crown 8vo, 300 pages.

Pleasant Half Hours; or, Thoughts for Men. By Rev. E. H. Sugden, M.A., Author of "Lights and Shadows," etc. With a Preface by Rev. A. Plummer, M.A., D.D., Master of University College, Durham. Crown 8vo, 224 pages.

Bible Picture Roll. Containing a large Engraving of a Scripture Subject, with letterpress, for each day in the month.

The Friends of Jesus: Illustrated Sketches for the Young, of the Twelve Apostles, the Family at Bethany, and other of the earthly friends of the Saviour. Small 4to. Cloth extra.

Love, Courtship, and Marriage. By Rev. F. B. Meyer, B.A. Crown 8vo, 152 pages. Embellished cloth cover. 2s. Net. Full gilt edges, 2s. 6d. Net.

2s. each *(continued)*.

By CHAS. M. SHELDON.

The Crucifixion of Phillip Strong. By Chas. M. Sheldon.

His Brother's Keeper. By Chas. M. Sheldon.

Our Exemplar; or, What would Jesus do? (In His Steps). By Chas. M. Sheldon.

Richard Bruce. By Chas. M. Sheldon.

The Twentieth Door. By Chas. M. Sheldon.

1s. 6d. each.

THRILLING BUT PURE!

A series of Charming Stories for Holiday and Fireside Reading.

Crown 8vo. 160 pages. Well Illustrated and Attractively Bound.

***A String of Pearls.** By E. F. Pollard.

***Elsie Macgregor:** or, Margaret's Little Lass. By Ramsay Guthrie.

The Lady of the Chine. By M. S. Haycraft.

Carola's Secret. By Ethel F. Heddle.

The Home of his Fathers. By Lillias Campbell Davidson.

A Great Patience. By L. Moberly.

In the Bonds of Silence. By J. L. Hornibrook.

A Late Repentance. By Hannah B. Mackenzie.

Shepherds and Sheep. By E. Stuart-Langford.

The Golden Doors. By M. S. Haycraft.

A Noble Champion. By David Hobbs.

THE UP-TO-DATE LIBRARY.

Of Thick Crown 8vo Volumes. 320 pages. Many Illustrations. Cloth boards.

***The Foster Brothers:** or, Foreshadowed. By Mrs. Morton.

***The Household Angel.** By Madeline Leslie.

The Green Mountain Boys: A Story of the American War of Independence. By E. F. Pollard.

A Way in the Wilderness. By Maggie Swan.

Miss Elizabeth's Niece. By M. S. Haycraft.

The Man of the House. By "Pansy."

1s. 6d. each *(continued).*

THE UP-TO-DATE LIBRARY *(continued)*

Olive Chauncey's Trust: A Story of Life's Turning Points. By Mrs. E. R. Pitman.

Whither Bound? A Story of Two Lost Boys. By Owen Landor.

Three People. By "Pansy."

Chrissy's Endeavour. By "Pansy."

The Young Moose Hunters. By C. A. Stephens.

Eaglehurst Towers. By Emma Marshall.

More Nails for Busy Workers. By C. Edwards. Author of "A Box of Nails for Busy Christian Workers," etc. Crown 8vo. 196 pages. Cloth boards.

Queen Alexandra: the Nation's Pride. By Mrs. C. N. Williamson. Crown 8vo. Tastefully bound. 1s. 6d. Net.

King and Emperor: The Life-History of Edward VII. By Arthur Mee. Crown 8vo. Cloth boards. 1s. 6d. Net.

William McKinley: Private and President. By Thos. Cox Meech. Crown 8vo. 160 pages, with Portrait. Net.

Studies of the Man Paul. By Robert E. Speer. Long 8vo. 304 pages. Cloth gilt. 1s. 6d. Net.

The Angel and the Demon; and other Stories. By E. Thorneycroft Fowler. Cloth gilt. Eight Illustrations.

A Measuring Eye. By E. Stuart-Langford, Author of "Miss Sophia's Repentance," etc. Illustrated. Cloth boards.

Wellington: The Record of a Great Military Career. By A. E. Knight. Crown 8vo. Cloth gilt, with Portrait. 1s. 6d. Net.

Hector Macdonald; or, The Private who became a General. By T. F. G. Coates. Cr. 8vo. Cloth gilt, with Portrait. 1s. 6d. Net.

Baden-Powell: The Hero of Mafeking. By W. Francis Aitken. Crown 8vo. Cloth gilt, with Portrait. 1s. 6d. Net.

Every-day Life in South Africa. By E. E. K. Lowndes. Crown 8vo. Illustrated. Cloth boards. 1s. 6d. Net.

The Royal Life. By Rev. J. C. Carlile. Crown 8vo. 128 pages. Cloth gilt.

Insects: Foes and Friends. By W. Egmont Kirby, M.D., F.L.S. 32 pages of Coloured Illustrations. Cloth boards.

The Romance of Evangelism. By Rev. J. Flanagan, South-East London Mission. Crown 8vo. 128 pages. Cloth boards.

1s. 6d. each *(continued).*

THE BRITISH BOYS' LIBRARY.

Fully Illustrated. Crown 8vo. 168 pages. Cloth extra.

Robin the Rebel. By H. Louisa Bedford.

Runaway Rollo. By E. M. Stooke.

Success: Chats about Boys who have Won it. By C. D. Michael.

Well Done! Stories of Brave Endeavour. Edited by C. D. Michael, Author of "Heroes All," "Deeds of Daring," etc.

The Wonder Seekers. By Henry J. Barker, M.A.

Little Soldiers. By Kate L. Mackley.

Will; or, That Boy from the Union. By Lydia Phillips.

Heroes All! A Book of Brave Deeds for British Boys. Edited by C. D. Michael.

Deeds of Daring; or, Stories of Heroism in Every-day Life. By C. D. Michael.

Noble Deeds: Stories of Peril and Heroism. Edited by C. D. Michael.

Armour Bright: The Story of a Boy's Battles. By Lucy Taylor, Author of "Astronomers and their Observations," etc.

The Thane of the Dean: A Story of the Time of the Conqueror. By Tom Bevan.

The Old Red School-house: A Story of the Backwoods. By Francis H. Wood.

Ben: A Story of Life's Byways. By Lydia Phillips.

The Secret of the Yew. By Frank Yerlock.

Major Brown; or, Whether White or Black, a Man. By Edith S. Davis.

The Bell Buoy; or, The Story of a Mysterious Key. By F. M. Holmes.

Jack. A Story of a Scapegrace. By E. M. Bryant.

Hubert Ellerdale: A Tale of the Days of Wicliffe. By W. Oak Rhind.

THE BRITISH GIRLS' LIBRARY.

Fully Illustrated. Crown 8vo. 160 pages. Cloth extra.

Granny's Girls. By M. B. Manwell.

Mousey; or, Cousin Robert's Treasure. By Eleanora H. Stooke.

1s. 6d. each *(continued).*

THE BRITISH GIRLS' LIBRARY *(continued).*

Marigold's Fancies. By L. E. Tiddeman.

"Our Phyllis." By M. S. Haycraft.

The Lady of Greyham; or, Low in a Low Place. By Emma E. Hornibrook.

The Gipsy Queen. By Emma Leslie.

Kathleen; or, A Maiden's Influence. By Julia Hack.

The Rajah's Daughter; or, The Half-Moon Girl. By Bessie Marchant.

In Self-Defence. By Julia Hack.

Regia: or, Her Little Kingdom. By E. M. Waterworth and Jennie Chappell.

Una's Marriage. By Mrs. Haycraft.

Tephi: An Armenian Romance. By Cecilia M. Blake.

Christabel's Influence. By J. Goldsmith Cooper.

Sweet Kitty Clare. By Jennie Chappell.

The Maid of the Storm: A Story of a Cornish Village. By Nellie Cornwall.

Queen of the Isles. By Jessie M. E. Saxby.

"THE WORLD'S WONDERS" SERIES.

Crown 8vo. 160 pages. Copiously Illustrated. Handsome cloth covers.

The Conquest of the Air: The Romance of Aerial Navigation. By John Alexander.

Surgeons and their Wonderful Discoveries. By F. M. Holmes.

The Life-Boat: Its History and Heroes. By F. M. Holmes.

Firemen and their Exploits. With an Account of Fire Brigades and Appliances. By F. M. Holmes.

Marvels of Ant Life. By W. F. Kirby, F.L.S., F.E.S., of the Natural History Museum, South Kensington.

The Romance of the Savings Banks. By Archibald G. Bowie.

The Romance of Glass-Making: A Sketch of the History of Ornamental Glass. By W. Gandy.

1s. 6d. each *(continued).*

"THE WORLD'S WONDERS" SERIES *(continued).*

The Romance of the Post-Office: Its Inception and Wondrous Development. By Archibald G. Bowie.

Marvels of Metals. By F. M. Holmes.

Miners and their Works Underground. By F. M. Holmes.

Triumphs of the Printing Press. By Walter Jerrold.

Astronomers and their Observations. By Lucy Taylor. With Preface by W. Thynne Lynn, B.A., F.R.A.S.

Celebrated Mechanics and their Achievements. By F. M. Holmes.

Chemists and their Wonders. By F. M. Holmes.

Engineers and their Triumphs. By F. M. Holmes.

Electricians and their Marvels. By Walter Jerrold.

Musicians and their Compositions. By J. R. Griffiths.

Naturalists and their Investigations. By George Day, F.R.M.S.

POPULAR MISSIONARY BIOGRAPHIES.

Crown 8vo. 160 pages. Cloth extra. Fully Illustrated.

James Chalmers, Missionary and Explorer of Rarotonga and New Guinea. By William Robson. New Edition, brought up-to-date by Frank B. Broad, of the London Missionary Society. With recent Portrait and many other Illustrations. 176 pages.

Griffith John, Founder of the Hankow Mission, Central China. By William Robson. New Edition, brought up-to-date by Frank B. Broad, of the London Missionary Society. 176 pages.

Robert Morrison: The Pioneer of Chinese Missions. By William J. Townsend.

Amid Greenland Snows; or, The Early History of Arctic Missions. By Jesse Page, F.R.G.S.

Bishop Patteson: The Martyr of Melanesia. By same Author.

Captain Allen Gardiner: Sailor and Saint. By same Author.

The Congo for Christ: The Story of the Congo Mission. By Rev. J. B. Myers, Author of "William Carey," etc.

David Brainerd, the Apostle to the North-American Indians. By Jesse Page, F.R.G.S.

1s. 6d. each *(continued).*

POPULAR MISSIONARY BIOGRAPHIES *(continued).*

David Livingstone: His Labours and his Legacy. By Arthur Montefiore-Brice, Author of "H. M. Stanley," etc.

From Kafir Kraal to Pulpit: The Story of Tiyo Soga, First Ordained Preacher of the Kafir Race. By Rev. H. T. Cousins.

Japan: Its People and Missions. By Jesse Page, F.R.G.S.

John Williams: The Martyr Missionary of Polynesia. By Rev. James Ellis.

James Calvert; or, From Dark to Dawn in Fiji. By R. Vernon.

Lady Missionaries in Foreign Lands. By Mrs. E. R. Pitman, Author of "Missionary Heroines in Eastern Lands."

Missionary Heroines in Eastern Lands. By Mrs. E. R. Pitman.

Reginald Heber: Bishop of Calcutta, Author of "From Greenland's Icy Mountains." By A. Montefiore-Brice, F.R.G.S.

Robert Moffat: The Missionary Hero of Kuruman. By David J. Deane.

Samuel Crowther: The Slave Boy who became Bishop of the Niger. By Jesse Page, F.R.G.S.

Thomas J. Comber: Missionary Pioneer to the Congo. By Rev. J. B. Myers.

William Carey: The Shoemaker who became the Father and Founder of Modern Missions. By Rev. J. B. Myers.

Henry Martyn. By Jesse Page, F.R.G.S.

NEW POPULAR BIOGRAPHIES.

Crown 8vo. Cloth boards. Amply Illustrated.

John Bright: Apostle of Free Trade. By Jesse Page, F.R.G.S.

The Two Stephensons. By John Alexander.

J. Passmore Edwards: Philanthropist. By E. Harcourt Burrage.

Dwight L. Moody: The Life-work of a Modern Evangelist. By Rev. J. H. Batt.

Noble Work by Noble Women: Sketches of the Lives of the Baroness Burdett-Coutts, Lady Henry Somerset, Miss Sarah Robinson, Mrs. Fawcett, and Mrs. Gladstone. By Jennie Chappell, Author of "Four Noble Women," etc.

1s. 6d. each *(continued).*

NEW POPULAR BIOGRAPHIES *(continued).*

Four Noble Women and their Work: Sketches of the Life and Work of Frances Willard, Agnes Weston, Sister Dora, and Catherine Booth. By Jennie Chappell.

The Canal Boy who became President. By Frederic T. Gammon.

Florence Nightingale: The Wounded Soldiers' Friend. By Eliza F. Pollard.

Four Heroes of India: Clive, Warren Hastings, Havelock, Lawrence. By F. M. Holmes.

Fridtjof Nansen: His Life and Explorations. By J. A. Bain.

General Gordon: The Christian Soldier and Hero. By G. Barnett Smith.

W. E. Gladstone: England's Great Commoner. By Walter Jerrold. With Portrait and 38 other Illustrations.

Heroes and Heroines of the Scottish Covenanters. By J. Meldrum Dryerre, LL.B., F.R.G.S.

John Knox and the Scottish Reformation. By G. Barnett Smith.

Philip Melancthon: The Wittemberg Professor and Theologian of the Reformation. By David J. Deane.

Sir Richard Tangye ("One and All"): An Autobiography. With 21 Original Illustrations by Frank Hewitt. 192 pages.

Sir John Franklin and the Romance of the North-West Passage. By G. Barnett Smith.

The Slave and His Champions: Sketches of Granville Sharp, Thomas Clarkson, William Wilberforce, and Sir T. F. Buxton. By C. D. Michael.

Henry M. Stanley: The African Explorer. By Arthur Montefiore-Brice, F.G.S., F.R.G.S.

C. H. Spurgeon: His Life and Ministry. By Jesse Page.

Two Noble Lives: John Wicliffe, the Morning Star of the Reformation; and Martin Luther, the Reformer. By David J. Deane. 208 pages.

William Tyndale: The Translator of the English Bible. By G. Barnett Smith.

The Marquess of Salisbury: His Inherited Characteristics, Political Principles, and Personality. By W. F. Aitken. *Net.*

Frederick Temple, Archbishop of Canterbury. By W. Francis Aitken, Author of "Baden-Powell," etc. *Net.*

1s. 6d. each *(continued).*

NEW POPULAR BIOGRAPHIES *(continued).*

Joseph Parker, D.D.: His Life and Ministry. By Albert Dawson. *Net.*

Hugh Price Hughes. By Rev. J. Gregory Mantle. *Net.*

NEW CENTURY LEADERS.

An up-to-date Series of Biographies of Men of Mark at the Opening of the Twentieth Century.

Crown 8vo. 160 pages and Portrait. Handsomely bound in cloth boards. 1s. 6d. each. Net.

R. J. Campbell, M.A.; Minister of the City Temple, London. By Charles T. Bateman.

Dr. Barnardo: "The Foster-Father of Nobody's Children." By Rev. J. H. Batt.

W. Robertson Nicoll, LL.D., Editor and Preacher. By Jane Stoddart.

F. B. Meyer: His Life and Work. By Jennie Street.

John Clifford, M.A., B.Sc., LL.D., D.D. By Chas. T. Bateman.

Canon Barnett, of the Toynbee Settlement. By W. Francis Aitken.

Alexander Maclaren, D.D.: The Man and His Message. By Rev. John C. Carlile.

Lord Milner. By W. B. Luke.

Lord Rosebery: Imperialist. By J. A. Hammerton, Author of "J. M. Barrie and His Books."

Joseph Chamberlain: A Romance of Modern Politics. By Arthur Mee.

General Booth: The Man and His Work. By Jesse Page.

ILLUSTRATED REWARD BOOKS.

Crown 8vo. 160 pages. Cloth extra. Fully Illustrated.

***Bethesda Chapel.** A Story of the Good Old Days. By Rev. C. Leach, D.D.

Philip's Inheritance; or, Into a Far Country. By F. Spenser.

Donald's Victory. By Lydia Phillips.

A Red Brick Cottage. By Lady Hope, Author of "His Handiwork," etc.

1s. 6d. each *(continued).*

ILLUSTRATED REWARD BOOKS *(continued).*

Marchester Stories. By Rev. C. Herbert.

Aileen; or, The Love of Christ Constraineth Us. By Laura A. Barter, Author of "Harold; or, Two Died for Me."

Everybody's Friend; or, Hilda Danvers' Influence. By Evelyn Everett-Green, Author of "Barbara's Brother," etc.

In Friendship's Name. A Story for Boys. By Lydia Phillips, Author of "Frank Burleigh, etc.

The Legend of the Silver Cup. Allegories for Children. By Rev. G. Critchley, B.A. With 12 Illustrations. Small 4to.

Nella; or, Not My Own. By Jessie Goldsmith Cooper.

Satisfied. By Catherine M. Trowbridge.

Sister Royal. By Mrs. Haycraft, Author of "The Children of Cherryholme," etc.

Ted's Trust; or, Aunt Elmerley's Umbrella. By Jennie Chappell, Author of "Who was the Culprit?" etc.

POPULAR EIGHTEENPENNY SERIES.

Crown 8vo. Well printed on good paper, and bound in attractive Cloth covers. Fully Illustrated.

Louie's Married Life. By Sarah Doudney.

Living it Down. By Laura M. Lane.

PICTURE BOOKS.

Size 9 by 7 inches. Coloured and numerous other Illustrations. Handsome Coloured Cover, Paper Boards with Cloth Back.

Happy and Gay: Pictures and Stories for Every Day. By D. J. D., Author of "Stories of Animal Sagacity," etc.

Pleasures and Joys for Girls and Boys. By D. J. D., Author of "Stories of Animal Sagacity."

Anecdotes of Animals and Birds. By Uncle John.

Stories of Animal Sagacity. By D. J. D. A companion volume to "Anecdotes of Animals."

1s. 6d. each *(continued).*

"ONWARD" TEMPERANCE LIBRARY.

Crown 8vo. Illustrated. Cloth extra.

Dick's Chum. By Miss M. A. Paull.
"This book is well written and illustrated. It is just the book for boys."

We Girls. By Miss M. A. Paull.
"A capital book for girls—written by one who thoroughly understands them."

The Fortunes of Riverside. By S. Hocking.
"This is a well-told tale, beautifully illustrated, and just the kind of book for our young folks."

Blossom and Blight. By Miss M. A. Paull, Author of "Tim's Troubles," etc.

Manor House Mystery. By Mrs. C. L. Balfour.
"It is written in excellent style, with a well-constructed plot, sparkling dialogue, and a faultless moral."

The Bird Angel. By Miss M. A. Paull.
"One of Miss Paull's most delightful stories."

Running from Home. By Miss M. A. Paull.
"A capital book for boys."

Lyndon the Outcast. By Mrs. Clara Lucas Balfour.

Ronald Clayton's Mistake. By Miss M. A. Paull.
"It is a capital book to place in the hands of working lads."

Nearly Lost, but Dearly Won. By Rev. T. P. Wilson, M.A., Author of "Frank Oldfield," etc.

Saph's Foster-Bairn. By Rev. A. Colbeck, Author of the £100 prize tale, "Fall of the Staincliffes," etc.

Hoyle' Popular Ballads and Recitations. By William Hoyle, Author of "Hymns and Songs," etc.
"A capital book for Sunday School, Temperance, and general Recitations."

1s. each.

"ONWARD" TEMPERANCE LIBRARY.

Crown 8vo. Illustrated. Cloth extra.

A Western Waif. By Old Cornish, Author of "Ste," "Our Girls," "Pete and his Daddy," etc., etc.

Addy's Two Lives. By Mrs. Ruth B. Yates, Author of "Grumpy Grafton," "Green and Gold," etc., etc.

The Gambler's Daughter; or, John Dudley's Secret. By Edward Armytage.

1s. each *(continued).*

"ONWARD" TEMPERANCE LIBRARY *(continued).*

The Tenants of Johnson's Court. By Janet Armytage.

Suspected; or, Under a Cloud. By A. J. Glasspool, Author of "Snatched from Death," etc.

Whispers to those who wish to Enjoy a Happy Life. By Rev. Benj. Smith, Author of "Climbing," "Gems Reset."

Snatched from Death. By Alfred J. Glasspool, Author of "The Young Abstainer's Laboratory," etc., etc.

From Cot to Crown. By Old Cornish, Author of "Ste," "Our Girls," "Pete and his Daddy," etc., etc.

1s. each.

ONE SHILLING REWARD BOOKS.

Fully Illustrated. Crown 8vo. Cloth extra.

Ruth's Roses: or, What Some Girls Did. By Laura A. Barter-Snow. Four Illustrations.

In Paths of Peril: A Boy's Adventures in Nova Scotia. By J. Macdonald Oxley. Six Illustrations.

Pets and their Wild Cousins: New and True Stories of Animals. By Rev. J. Isabell, F.E.S. Twenty Illustrations.

A Brother's Need. By L. S. Mead. Crown 8vo. 128 pages. Illustrated. Cloth.

Sunshine and Snow. By Harold Bindloss. Six Illustrations.

Donalblane of Darien. By J. Macdonald Oxley. Six Illustrations.

Crown Jewels. By Heather Grey. Four Illustrations.

At the Bend of the Creek. By E. Gertrude and Annie A. Hart. Four Illustrations.

All Play and No Work. By Harold Avery, Author of "The Triple Alliance," etc. Six Illustrations by Harold Copping.

Bernard or Ben? By Jennie Chappell, Author of "Raymond's Rival," etc. Six Illustrations by T. L. Pethybridge.

Always Happy; or, The Story of Helen Keller. By Jennie Chappell, Author of "Ted's Trust."

Birdie and her Dog, and other Stories of Canine Sagacity. By Miss Phillips (Mrs. H. B. Looker).

1s. each *(continued).*

ONE SHILLING REWARD BOOKS *(continued).*

Bessie Drew; or, The Odd Little Girl. By Amy Manifold.

Cola Monti; or, The Story of a Genius. By Mrs. Craik, Author of "John Halifax, Gentleman."

The Children of Cherryholme. By M. S. Haycraft.

The Fatal Nugget. By E. Harcourt Burrage.

Frank Burleigh; or, Chosen to be a Soldier. By Lydia Phillips.

Harold; or, Two Died for Me. By Laura A. Barter.

Indian Life in the Great North-West. By Egerton R. Young, Missionary to the North American Indian Tribes.

Jack the Conqueror; or, Difficulties Overcome. By the Author of "Dick and his Donkey."

Jim's Discovery; or, On the Edge of a Desert. By T. M. Browne, Author of "Dawson's Madge," etc.

Little Bunch's Charge; or, True to Trust. By Nellie Cornwall, Author of "Tamsin Rosewarne," etc.

Lost in the Backwoods. By Edith C. Kenyon.

The Little Woodman and his Dog Cæsar. By Mrs. Sherwood.

The Last Look: A Tale of the Spanish Inquisition. By W. H. G. Kingston, Author of "Manco, the Peruvian Chief."

Marjory; or, What would Jesus do? By Laura A. Barter.

Our Den. By E. M. Waterworth, Author of "Master Lionel, that Tiresome Child."

Paul the Courageous. By Mabel Quiller-Couch. Six Illustrations by Florence Reason.

Roy's Sister; or, His Way and Hers. By M. B. Manwell.

Raymond's Angel: A Story of Two Lives Laid Down. By Blanche Garvock. Illustrated. Crown 8vo. Cloth boards.

Raymond's Rival; or, Which will Win? By Jennie Chappell.

St. Mary's Convent; or, Chapters in the Life of a Nun. By J. S. Dammast, Author of "The Fatal Legacy."

Sweet Nancy. By L. T. Meade, Author of "Scamp and I."

Uncle Zeph and His Yarns. By Wm. J. Forster, Author of "A Trip to Many Lands," etc. Four Illustrations.

Who was the Culprit? By Jennie Chappell.

1s. each *(continued.)*

ONE SHILLING BOOKS *(continued).*

The Armour of Life. A Little Book of Friendly Counsel. Edited by J. A. Hammerton, Author of "A Young Man's Mind." Foolscap 8vo. Ninety-six pages. Cloth.

Hiram Golf's Religion. By George H. Hepworth, D.D., Author of "The Life Beyond," etc. 128 pages. Cloth gilt.

Eon the Good; and other Verses. By Charlotte Murray. Crown 8vo.

What is Christian Science? An Examination of the Metaphysical, the Theological, and the Therapeutic Theories of the System. By P. C. Wolcott, B.D. Long 8vo. 96 pages. Cloth.

Another Pentecost. By Rev. I. E. Page. Long 8vo. 128 pages. Cloth boards.

Uncrowned Queens. By Charlotte Skinner, Author of "Sisters of the Master." Small 8vo. 112 pages. Cloth.

Sisters of the Master. By Charlotte Skinner, Author of "The Master's Gifts to Women."

Some Secrets of Christian Living: Selections from the "Seven Rule" Series of Booklets. Small 8vo. Cloth boards.

Light and Darkness; or, All One in Christ Jesus. By Spes. An Anti-Ritualistic Story. Crown 8vo. 128 pages. Paper covers.

Molly and I. By the Author of "Jack," "At Sunset," etc. Long 8vo. Illustrated Title Page.

Cicely's Little Minute. By Harvey Gobel. Long 8vo. Illustrated Title Page. Cloth extra.

Victoria: the Well-Beloved. (1819-1901.) By W. Francis Aitken, Author of "Baden Powell: the Hero of Mafeking." Eight Illustrations. Crown 8vo. 152 pages. Cloth boards.

CHEAP REPRINTS OF POPULAR BOOKS FOR THE YOUNG.

Crown 8vo. 160 pages. Illustrated. Cloth boards, 1s. each.

*__A Candle Lighted by the Lord.__ By Mrs. E. Ross.

*__Alice Western's Blessing.__ By Ruth Lamb.

Tamsin Rosewarne and Her Burdens: A Tale of Cornish Life. By Nellie Cornwall.

1s. each *(continued).*

CHEAP REPRINTS OF POPULAR BOOKS FOR THE YOUNG *(contd).*

Raymond and Bertha: A Story of True Nobility. By Lydia Phillips, Author of "Frank Burleigh; or, Chosen to be a Soldier."

Gerald's Dilemma. By Emma Leslie. Crown 8vo. 160 pages. Six Illustrations. Cloth.

Fine Gold; or, Ravenswood Courtenay. By Emma Marshall, Author of "Eaglehurst Towers," etc.

Marigold. By Mrs. L. T. Meade, Author of "The Little Princess of Tower Hill," etc.

Jack's Heroism. A Story of Schoolboy Life. By Edith C. Kenyon.

The Lads of Kingston. A Tale of a Seaport Town. By James Capes Story.

Her Two Sons: A Story for Young Men and Maidens. By Mrs. Charles Garnett.

Rag and Tag: A Plea for the Waifs and Strays of Old England. By Mrs. E. J. Whittaker.

Through Life's Shadows. By Eliza F. Pollard, Author of "Roger the Ranger."

The Little Princess of Tower Hill. By L. T. Meade, Author of "The Lady of the Forest."

Clovie and Madge. By Mrs. G. S. Reaney.

Ellerslie House: A Book for Boys. By Emma Leslie.

Like a Little Candle; or, Bertrand's Influence. By Mrs. Haycraft.

Martin Redfern's Vow. By Ethel F. Heddle.

The Dairyman's Daughter. By Legh Richmond.

Bible Wonders. By Rev. Dr. Newton.

The Pilgrim's Progress. By John Bunyan. 416 pages. Forty-seven Illustrations.

Our Duty to Animals. By Mrs. C. Bray, Author of "Physiology for Schools," etc. Intended to teach the young kindness to Animals. Cloth, 1s. 6d.; School Edition, 1s. 3d.

1s. each *(continued).*

NEW SERIES OF ONE SHILLING PICTURE BOOKS.

Size 10½ by 8 inches. 96 pages. Coloured Frontispiece and numerous other illustrations. Handsomely bound in paper boards, covers printed in 10 colours and varnished.

Two Little Bears at School. By J. D., Author of "Happy Playmates," "Buttercups and Daisies," etc.

Merry and Free. Pictures and Stories for our Little Ones. By R. V.

We three and Grandpa! A Picture Story Book for Little Folks. By J. D.

Bruno's Prize. Pictures and Stories for Merry and Wise. By D. J. D., Author of "Our Pet's Picture Book."

Our Pet's Picture Book. By D. J. D.

Happy Playmates: Pictures and Stories for Little Folks. By J. D., Author of "Buttercups and Daisies," etc.

Bible Pictures and Stories: Old Testament. By D. J. D.

Bible Pictures and Stories: New Testament. By James Weston and D. J. D.

Pussies and Puppies. By Louis Wain.

Dapple and Dobbin's Picture Book. By D. J. D.

Foolscap 4to. With Coloured Covers, and full of Illustrations.

Buttercups and Daisies: A Picture Story Book for Little People. By J. D., Author of "Ring o' Roses," etc. Size 9 by 7½ inches. One full-page coloured and many other Illustrations.

Holiday Hours in Animal Land. By Uncle Harry. New edition. 96 pages. One full-page Coloured and numerous other Illustrations. Paper boards with coloured design.

Ring o' Roses. Pictures and Stories for Little Folks. By Uncle Jack, Author of "Frolic and Fun," etc. Four full-page Coloured and numerous other Illustrations.

Holiday Joys. Stories and Pictures for Girls and Boys. By C. D. M., Author of "Merry Playmates, etc." Four full-page Coloured and numerous other Illustrations.

Commendations from all parts of the world have reached Messrs. S. W. Partridge and Co. upon the excellence of their Picture Books. The reading matter is high-toned, helpful, and amusing, exactly adapted to the requirements of young folks; while the Illustrations are by first-class Artists, and the paper is thick and durable. Bound in attractive coloured covers, they form a unique series.

1s. each *(continued).*

ONE SHILLING TOY BOOKS.

Father Time. A Novel Mechanical Toy Book. Cover tastefully printed in colours. Contains beautifully reproduced illustrations of the games played during each month of the year; also clock-dial with real hands.

Animals, Tame and Wild. Fourteen coloured pages of Animals drawn from life, with appropriate footlines. Beautifully coloured cover, varnished.

DEVOTIONAL BOOKS FOR CHRISTIAN WORKERS.

Large Crown 16mo. 128 pages. Chastely bound in Cloth Boards. 1s. each.

Some Deeper Things. By Rev. F. B. Meyer.

Steps to the Blessed Life. By Rev. F. B. Meyer.

Daybreak in the Soul. By Rev. E. W. Moore.

The Temptation of Christ. By C. Arnold Healing, M.A.

Keynotes to the Happy Life. By Charlotte Skinner, Author of "The Master's Gifts to Women," etc.

For Love's Sake. By Charlotte Skinner, Author of "Uncrowned Queens," etc.

9d. each.

NINEPENNY SERIES OF ILLUSTRATED BOOKS.

96 pages. Small Crown 8vo. Illustrated. Handsome cloth covers.

Grandmother's Child. By Annie S. Swan. New Edition.

Dorothy's Trust. By Adela Frances Mount, Author of "Margery's Quest."

Grannie's Treasures; and how they helped her. By L. E. Tiddeman.

His Majesty's Beggars. By Mary E. Ropes.

Love's Golden Key. By Mary E. Lester.

Faithful Friends. By C. A. Mercer, Author of "Rob and I."

Only Roy. By E. M. Waterworth and Jennie Chappell.

Aunt Armstrong's Money. By Jennie Chappell, Author of "Carol's Gift," etc.

9d. each *(continued).*

NINEPENNY SERIES OF ILLUSTRATED BOOKS *(continued).*

The Babes in the Basket; or, Daph and Her Charge.

Bel's Baby. By Mary E. Ropes.

Birdie's Benefits; or, A Little Child Shall Lead Them. By Edith Ruth Boddy.

Carol's Gift; or, "What Time I am Afraid I will Trust in Thee." By Jennie Chappell, Author of "Without a Thought."

Cripple George; or, God has a Plan for Every Man. A Temperance Story. By John W. Kneeshaw.

Cared For; or, The Orphan Wanderers. By Mrs. C. E. Bowen, Author of "Dick and his Donkey," etc.

Dawson's Madge: or, The Poacher's Daughter. By T. M. Browne, Author of "The Musgrove Ranch," etc.

A Flight with the Swallows. By Emma Marshall.

The Five Cousins. By Emma Leslie.

Foolish Chrissy; or, Discontent and its Consequences. By Meta, Author of "Noel's Lessons," etc.

For Lucy's Sake. By Annie S. Swan.

Giddie Garland; or, The Three Mirrors. By Jennie Chappell.

How a Farthing Made a Fortune; or, Honesty is the Best Policy. By Mrs. C. E. Bowen.

How Paul's Penny became a Pound. By Mrs. Bowen, Author of "Dick and his Donkey."

How Peter's Pound became a Penny. By the same Author.

John Blessington's Enemy: A Story of Life in South Africa. By E. Harcourt Burrage, Author of "The Fatal Nugget," etc.

John Oriel's Start in Life. By Mary Howitt.

Master Lionel, that Tiresome Child. By E. M. Waterworth.

The Man of the Family. By Jennie Chappell.

Mattie's Home; or, The Little Match-girl and her Friends.

Nan; or, The Power of Love. By Eliza F. Pollard, Author of "Avice," etc.

Phil's Frolic. By F. Scarlett Potter.

9d. each *(continued).*

NINEPENNY SERIES OF ILLUSTRATED BOOKS *(continued).*

Paul: A Little Mediator. By Maude M. Butler.

Rob and I; or, By Courage and Faith. By C. A. Mercer.

A Sailor's Lass. By Emma Leslie.

Una Bruce's Troubles. By Alice Price.

Won from the Sea. By E. C. Phillips (Mrs. H. B. Looker), Author of "Birdie and Her Dog."

6d. each.

NEW SERIES OF SIXPENNY PICTURE BOOKS.

Sweet Stories Retold: A Bible Picture Book for Young Folks. Crown 4to. Coloured Frontispiece and 25 Illustrations. Handsomely bound in paper boards, with Cover printed in Ten Colours.

Little Snowdrop's Bible Picture Book. Crown 4to. Coloured Frontispiece and 25 Illustrations. Handsomely bound in paper boards, with cover printed in Ten Colours.

March Away! Pictures and Stories for Every Day. Coloured Frontispiece and many other Illustrations. Handsomely bound in paper boards, with Cover printed in Ten Colours.

After the Ball: Pictures and Stories for One and All. Coloured Frontispiece and many other Illustrations. Handsomely bound in paper boards, with cover printed in Ten Colours.

Sweet Blossom: A Picture Story Book for Little Ones. Crown 4to. Fully Illustrated. Handsomely bound in paper boards, with design printed in eight colours.

THE MARIGOLD SERIES.

An entirely new and unequalled series of Standard stories, printed on good laid paper. Imperial 8vo. 128 pages. Illustrated covers with vignetted design printed in eight Colours. Price 6d. each net.

Pride and Prejudice. By Jane Austen.

From Jest to Earnest. By E. P. Roe.

The Wide, Wide World. By Susan Warner.

POPULAR EDITION.

Don Lavington; or, In the Days of the Press Gang. By Geo. Manville Fenn. Royal 8vo. 160 pages. Ten Illustrations by W. Rainey, R.I. Cover printed in colours.

6d. each *(continued)*.

THE "RED DAVE" SERIES.

New and enlarged Edition. Handsomely bound in cloth boards. Well Illustrated.

Wilful Jack. By M. I. Hurrell.

Willie the Waif. By Minie Herbert.

A Sunday Trip and what came of it. By E. J. Romanes.

Little Tim and his Picture. By Beatrice Way.

Midge. By L. E. Tiddeman.

The Conjuror's Wand. By Henrietta S. Streatfeild.

Benjamin's New Boy.

Enemies: A Tale for Little Lads and Lasses.

Cherry Tree Place.

A Tale of Four Foxes.

A Little Town Mouse.

The Little Governess.

Puppy-Dog Tales.

Mother's Boy.

A Great Mistake.

From Hand to Hand.

That Boy Bob.

Buy Your Own Cherries.

Left in Charge, and other Stories

A Threefold Promise.

The Four Young Musicians.

Two Little Girls and What they did.

Joe and Sally; or, A Good Deed and its Fruits.

The Island Home.

Chrissy's Treasure.

Lost in the Snow.

Owen's Fortune.

Red Dave: or, What Wilt Thou have Me to Do.

Dick and His Donkey.

Jessie Dyson.

Come Home, Mother.

4d. each.

CHEAP "PANSY" SERIES.

Imperial 8vo. 64 pages. Many Illustrations. Cover printed in five colours

The Strait Gate. By Annie S. Swan.

Mark Desborough's Vow. By Annie S. Swan.

Her Saddest Blessing.

Miss Priscilla Hunter, and other Stories.

Wild Bryonie.

Avice. A Story of Imperial Rome.

Links in Rebecca's Life.

From Different Standpoints.

Those Boys.

Christie's Christmas.

Four Girls at Chautauqua.

Ester Ried.

Julia Ried.

Ester Ried yet Speaking.

Echoing and Re-echoing.

Cunning Workmen.

Tip Lewis and His Lamp.

The King's Daughter.

Household Puzzles.

The Randolphs.

Wise to Win; or, The Master Hand.

A New Graft on the Family Tree.

The Man of the House.

4d. each (continued).

THE YOUNG FOLKS' LIBRARY,

of Cloth-bound Books. With Coloured Frontispiece. 64 pages. Well Illustrated. Handsome Cloth Covers.

The Little Woodman.
Jacko the Monkey, and other Stories.
Little Dan, the Orange Boy.
Ronald's Reason.
From Shadow to Sunshine.
A Bright Idea.
Sybil and her Live Snowball.
The Church Mouse.
Dandy Jim.
A Troublesome Trio.
Perry's Pilgrimage.
Nita ; or, Among the Brigands.

3d. each.

NEW PRETTY "GIFT-BOOK" SERIES.

With Beautiful Coloured Frontispiece, and many other Illustrations. Paper boards, Cover printed in eight Colours and Varnished, 3d. Size 6 by 5 inches.

Jack and Jill's Picture Book.
Lady - Bird's Pictures and Stories.
Playtime Joys for Girls and Boys.
Dolly's Picture Book.
By the Sea.
Toby and Kit's Animal Book.
"Pets" and "Pickles"
Our Little Pets' Alphabet.
Bible Stories—Old Testament.
Bible Stories—New Testament.

THE TINY LIBRARY.

Books printed in large type. Cloth limp, 3d.

Little Chrissie, and other Stories.
Harry Carlton's Holiday.
A Little Loss and a Big Find.
What a Little Cripple Did.
Bobby.
Matty and Tom.
The Broken Window.
John Madge's Cure for Selfishness.
The Pedlar's Loan.
Letty Young's Trials.
Brave Boys.
Little Jim, the Rag Merchant.

PARTRIDGE'S PICTORIAL MAGAZINES.

"A word of emphatic praise should be given to the old established and excellent magazines of Messrs. Partridge and Co. They ought to hold their own against the great competition, for they are eminently sound, healthy, and interesting."—BRITISH WEEKLY.

"It would be difficult to surpass these magazines. All have marched with the times." DAILY TELEGRAPH.

"There are no more attractive Annual Volumes than those issued by S. W. Partridge and Co."—The CHRISTIAN.

Our Young Men. A Magazine of New Century Ideas. Edited by J. A. Hammerton. A thoroughly bright, breezy, go-ahead magazine, with attractive articles, serial stories, answers to correspondents, book chat, etc., etc. 1d. Monthly.

The Yearly Volume, cloth boards, **2s.**

The British Workman. Contains Popular Articles and Stories on Temperance, Thrift, etc., and short Biographies of eminent Self-made Men; also interesting accounts of visits to some of our leading British Industries. 1d. Monthly.

The Yearly Volume, 96 pages full of illustrations, coloured paper boards, **1s. 6d.**; cloth, **2s. 6d.**

The Family Friend. A beautifully Illustrated Magazine for the Home Circle, with Serial and Short Stories by popular Authors, Helpful Articles, Hints on Dressmaking, Music, etc. 1d. Monthly.

The Yearly Volume, in coloured paper boards and cloth back, **1s. 6d.**; cloth, **2s.**; gilt edges, **2s. 6d.**

The Friendly Visitor. An Illustrated Gospel Magazine for the people, full of entertaining reading with sound religious teaching in the form of story, article, and poem. Printed in large type and fully illustrated. Just the paper for the aged. 1d. Monthly.

The Yearly Volume, coloured paper boards and cloth back, **1s. 6d.**; cloth, **2s.**; gilt edges, **2s. 6d.**

The Children's Friend. Charming Stories, interesting Articles, Indoor Recreations, beautiful Pictures, Puzzles, Music, Prize Competitions, etc. 1d. Monthly.

The Yearly Volume, coloured paper boards, with cloth back and excellent coloured frontispiece, **1s. 6d.**; cloth, **2s.**; gilt edges, **2s. 6d.**

The Infants' Magazine. No other Periodical can be compared with the Infants' Magazine for freshness, brightness, and interest. Full of bright pictures and pleasant reading to delight the little ones. 1d. Monthly.

The Yearly Volume, coloured paper boards, with cloth back and beautifully coloured frontispiece, **1s. 6d.**; cloth, **2s.**; gilt edges, **2s. 6d.**

The Band of Hope Review. The Leading Temperance Periodical for the Young, containing Serial and Short Stories, Concerted Recitations, Prize Competitions. Should be in the hands of all Members of Bands of Hope. ½d. Monthly.

The Yearly Volume, coloured paper boards, **1s.**; cloth boards **1s. 6d.**

Printed in Dunstable, United Kingdom

91222427R00221